Reviews of *Services for UMTS:*

"In the last few years we have heard fantastic things about 3rd generation systems and the incredible services they will provide. Unfortunately most of these were just that: fantastic and incredible. On the other hand most of us missed the point of what the next generation mobile can enable and what real new services are becoming possible. This book is a must read if you want to understand options, future services and dream about them from a rock solid standpoint"

Roberto Saracco, Director, Future Centre, Telecom Italia Lab

"This book is a visionary outlook into the world of UMTS and its compelling services. It outlines how modern tools can be used in mobile marketing to add value and utility to the user"

Andreas X. Müller, Executive Board, 12Snap AG

"This is certainly the most comprehensive work I have seen on the subject. The book explains how various elements of technology, product development and system integration have come together to build successful 3G services"

Regina Nilsson, Director, Telecom Practice, PwC Consulting, Northern Business Unit

"The authors provide an insightful discussion into a wealth of service possibilities that could be delivered by UMTS. This will potentially offer significant revenue opportunities and bring values to mobile operators and may also enable service enhancement with existing access technologies"

Dr. Stanley Chia, Director, Group R&D - US, Vodafone

"In this book the editors succeed at building a better understanding of UMTS. This should help telecom operators, equipment manufacturers, content providers and the capital markets manage their $1 trillion bet on the success of 3G"

Assaad Razzouk, Deputy Head Global Corp Finance, Nomura International plc

"A welcome change from the technology-led literature, *Services for UMTS* focuses on the services and applications end of the mobile multimedia world. Through an interesting framework the editors have managed to explain how value can be created from both a user and a service provider perspective"

Dr Didier Bonnet, Global Head of Strategic and Business Consulting,
Telecom and Media Practice, Cap Gemini Ernst & Young

"This book explains some of the compelling services the players in the wireless industry will be able to develop and deploy based on the 3G and 4G infrastructure"

Jeff Lawrence, Director of Technology, Intel

"Services that customers need will be the only driver for 3G. This book provides a framework for the launch of UMTS, but more significantly strong ideas for future demand and capability"

Mike Short, Vice President mmO2, Past Chairman GSM Association

SERVICES FOR UMTS

SERVICES FOR UMTS

CREATING KILLER APPLICATIONS IN 3G

Edited by

Tomi T Ahonen
Independent Consultant, UK

Joe Barrett
Nokia Networks, Finland

JOHN WILEY & SONS, LTD

Baffins Lane, Chichester,
West Sussex, P019 IUD, England

National 01243 779777
International (+44)1243 779777

e-mail (for orders and customer service enquiries): cs-books@wiley.co.uk

Visit our Home Page on http://www.wileyeurope.com or http://www.wiley.com

Other Wiley Editorial Offices

John Wiley & Sons, Inc., 605 Third Avenue,
New York, NY 10158-0012, USA

WILEY-VCH Verlag GmbH
Pappelallee 3, D-69469 Weinheim, Germany

John Wiley & Sons Australia Ltd, 33 Park Road, Milton,
Queensland 4064, Australia

John Wiley & Sons (Canada) Ltd, 22 Worcester Road
Rexdale, Ontario, M9W lLl, Canada

John Wiley & Sons (Asia) Pte Ltd, 2 Clementi Loop #02-01,
Jin Xing Distripark, Singapore 129809

A catalogue record for this book is available from the British Library

ISBN 0471 485500

Typeset in Sabon by Deerpark Publishing Services Ltd, Shannon, Ireland.
Printed and bound in Great Britain by T. J. International Ltd, Padstow, Cornwall.

This book is printed on acid-free paper responsibly manufactured from sustainable forestry, in which at least two trees are planted for each one used for paper production.

Images on cover were supplied by Nokia Corporate Communications, Finland

Contents

Foreword

One of the fundamental questions being asked about the 3rd Generation (3G) or UMTS business is 'Will it be a profitable business and a viable investment?' There are many opinions and views on this subject but only time will reveal the answer. Of course it is easy to say that UMTS will be a resounding success. It is even easier to take a marketing view of the world where theory is the basis for your UMTS messages and communicating services more relevant to 2010 are your marketing objective. We have to ensure we are not victims of our own technological hype and enthusiasm for the future. If we over sell the technology and increase user expectations beyond the realms of reality we run the risk of alienating the customer and failing in the delivery of a totally new UMTS user experience.

There are few successful revolutions in our age and UMTS will not change our lives over night. What it will do is bring a new way to interact with people, devices, information and businesses. UMTS will change our lives but gradually, not over 40 years, 20 years or even 10 years which has been the past norm but in a shorter rhythm, an Internet time span that is measured in tens of months not tens of years. This speed brings with it the need for operators to be responsive, adaptive and nimble.

The Internet has changed our lives for the better. Consumers have more power than they had only a few years ago and this is good for industry since it creates a competitive environment where price, quality and service are the winning combination. We see this in the success of companies that provide more personalised services via electronic access to content. However, it is the success and growth of Mobility that will have the greatest impact on how we communicate and interact with society. If Content is King in the Internet, it will be Personalised Content that is relevant, timely and localized for the user's situation that will be King in the Mobile Internet.

The current Internet has been a victim of its own success. The euphoria of the early years is fading slowly as click through rates fall and advertisers re-evaluate the effectiveness of the Internet business model. At the heart of the problem is invoiceability. Without end-to-end control over the transaction there is no validation of the value that each user contact generates. In effect the Internet has created the perception that there is such a thing as a "free lunch" or in some cases even a "profitable lunch".

Serious readers know better. It will be mobile users that create new demands on content. Mobility requirements will change the Internet into a content provisioning environment that creates real value. Value that consumers and executives will pay for. Our industry has to create a "win-win" situation between subscribers, operators, content providers, developers and vendors. New partnerships are critical, in which new entrants can prosper.

I believe that 3G is fundamentally a good business, and exciting. Market acceptance of 3G services on a global scale is within reach. Key words are multimedia, ubiquity, simplicity, affordability, and globalization. The successful companies will understand the changes in user expectations and they will meet them and exceed them. The organizations that do that will own the future. They and their shareholders will be the long term winners.

Alan Hadden
President, GSA
Global mobile Suppliers Association
www.gsacom.com

'I love deadlines. I like the whooshing sound they make as they fly by.'

Douglas Adams

Acknowledgements

Our guides, mentors, advisors and gurus

A book like this would not have been possible without the valuable input of our contributors and the many others who helped and inspired us.

We want to take this opportunity to thank a few very special people who have advised us, guided us and in some cases who were kind enough to take time to provide input and critique for this book when it was still in manuscript form.

Among our personal mentors and advisors in understanding the very nature of UMTS have been Ukko Lappalainen, Ilkka Pukkila and Ebba Dåhli. For their visions, foresight and guidance we are very grateful. In the areas of econometric modelling and understanding the operator business case for UMTS, we want to thank Hannu Tarkkanen, Timo M Partanen, Paulo Puppoli, Vesa Sallinen, Petro Airas and Harri Leiviskä. In areas of mobile services and their revenues we are very grateful to Claus von Bonsdorff, Nicole Cham, Heikki Koivu, Michael Addison, Timo Kotilainen, and Timo Poikolainen. In understanding the business customer needs of UMTS we thank Julian Heaton, and in residential customer needs Reza Chady and Paul Bloomfield. In helping us understand UMTS operator needs we thank Merja Kaarre, Carina Lindblad, Jaakko Hattula and Spencer Rigler. We want to remember Tarmo Honkaranta for his leadership in developing and promoting the use of segmentation.

Several visionaries inspired us and specifically we feel a debt of gratitude to Teppo Turkki, Matti Makkonen, Risto Linturi, Taina Kalliokoski, Voytek Siewierski, and Sakuya Morimoto. Both of us have found considerable insight into possible technological future scenarios in Scott Adams's book 'The Dilbert Future'.

We are very grateful for the patience and guidance given by John Wiley & Sons, Ltd, especially Mark Hammond. Thank you for your endless patience and steadfast support.

For their patience, understanding, support and encouragement we are truly grateful to our families and friends for the weekends and late nights devoted to this book. Special thanks go to Kay Barrett who read every page of the manuscript and often made sense out of our ideas and text.

It has been a challenge to live by the tight rules we imposed upon our contributors and ourselves and perhaps we can all empathise with Peter de Vries who said "I love being a writer. What I can't stand is the paperwork."

We welcome comments or suggestions for improvements or changes that will improve future editions of this book. The e-mail address for suggestions is s4umts@hotmail.com.

Tomi T Ahonen and Joe Barrett

List of Contributors

Editors

Tomi T Ahonen is an independent consultant specialising in 3G Strategy at www.tomiahonen.com.

Joe Barrett works in Nokia Networks Customer Marketing focusing on a variety of 2.5G and 3G topics.

Both Tomi and Joe co-edited the book and provided contributions to many of the chapters.

Contributors

Jouko Ahvenainen works in telecoms market segmentation and tools for it at segmentation specialist firm Xtract Ltd. Jouko contributed to the chapters on Marketing, Competitiveness and Partnering.

Russell Anderson is a Launch Marketing Manager, working to bring Nokia's new terminals to the market place. Russell contributed to the chapter on Services to the Me attribute.

Frank Ereth is a Director of Sales and Marketing for Mobile Applications at Wanova. Frank contributed to the chapter on Partnering.

Harri Holma is a System Expert with 3G radio networks at Nokia with special interest in radio network performance and optimisation. Harri is the co-editor of 'WCDMA for UMTS' and contributed to the Technical Primer chapter.

Päivi Keskinen is a Business Consultant at HiQ, a telecom consulting company. Päivi contributed to the chapters on Movement and Moment services.

Ari Lehtoranta works for Nokia where he recently headed the global sales and delivery organisation for Nokia's Mobile Internet Applications. Ari contributed to the chapters on Types of Services and Competitiveness.

Canice Mckee is a 3G expert with experience in operator business and markets. Canice contributed to the chapter on Competitiveness.

Timo Rastas is a 3G expert at Nokia. Timo contributed to the chapter on Marketing.

Michael D Smith is the Director of Services with APSolve (www.apsolve.com) a company specialising in scheduling and management of large remote workforces. Michael contributed to the chapters on Movement services and Marketing.

Mika Suomela is a Senior Marketing Manager at Technology Marketing, Nokia Networks, focusing on future end-user services and service enablers. Mika contributed to the chapter on Services to the Me attribute.

Antti Toskala a working as a Standardisation Manager at Nokia Networks and he is currently involved in WCDMA and GSM standardisation at 3GPP. Antti is the co-editor of 'WCDMA for UMTS' and contributed to the Technical Primer chapter.

1

'No one would ever have crossed the ocean if he could have gotten off the ship in the storm.'
Charles Kettering

Intro to Services for UMTS:
The Future Starts Here

Joe Barrett and *Tomi T Ahonen*

Kotler, Porter and other eminent marketing gurus have preached that the first step in the marketing cycle is to segment your market. From there you position your product and then target the audience with the right messages. Traditionally this was done by social groupings, A, B, C1:C2:C3 or other demographic methods. Since those early days marketers have been seeking and developing new segmentation strategies and many have been used to good effect, but the global trend is towards ever smaller and more precise segments, approaching the ideal segment of one. A segment of one means that a marketer can target each individual on a one-to-one basis and has the greatest opportunity to take the potential customer through the buying cycle: Awareness, Interest, Decision, Action.

Previously on a practical level it has been almost impossible to segment your target audience in this way, primarily because of cost reasons. Today this is changing with the Internet, where target advertising can be used with reasonable results. If you are accessing a golf page for example you are more likely to see a banner advert that is golf or maybe sport related. However in UMTS (Universal Mobile Telecommunications System) networks this will change. In the near future, ever tighter segments and more precision in market messages becomes not only possible, but necessary for successful profits.

We are all individuals. We come in different sizes, shapes, colours. We have different needs, desires, wants. We all do things in our own unique way. It is the fact that we are all unique and different that unites us. Once we recognise this we can start to exploit it in our marketing. As soon as we can build an individual relationship with our customers, when we know what they want, what they need, how they do things, run their work life, personal life and how they manage their relationships we can show them how they can make their lives easier, more profitable and more fun. Sounds too good to be true? Not if you are a mobile phone operator. Voice has already gone wireless and data is the new frontier.

1.1 Enriching the experience. From ears to eyes

At the heart of this UMTS experience will be the terminal and a new way of using the phone. The mobile subscriber will not just talk, they will be able to view multimedia images, watch video clips, listen to music, shop, book a restaurant table and surf the net. And, since they will always be connected to the network, they will receive important and timely information.

The strong growth in mobile voice will continue in mobile data. There were around 630 million mobile phone users in 2000 and this number is expected to grow to 1 billion by the end of 2002[1]. For comparison there are less than 300 million personal computers in the world, connected to the Internet. The UMTS terminal will become a service platform, capable of multiple radio access modes and compliant with *open standards and operating systems*[2] to enable *Mobile Internet* and *mobile Multimedia Messaging Services*.

[1] Strategy Analytics' February 2000 report, 'Worldwide Cellular Markets 2000–2005'.
[2] The open standards supported by Nokia terminals include WAP; Bluetooth; EPOC; SyncML; HSCSD, GPRS, EDGE and WCDMA.

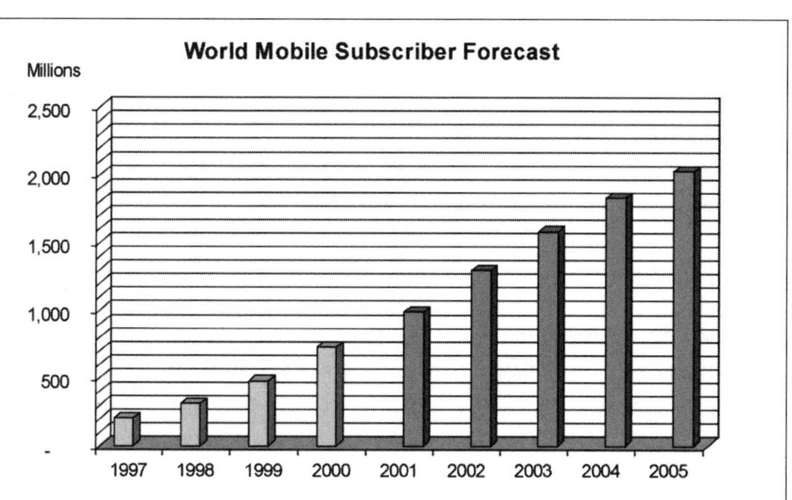

The growth of mobile subscribers has been remarkable over the past 10 years. This has been driven mostly from Europe and Asia where GSM has been the dominant technology. The adoption of WCDMA by operators in Europe, Asia, USA, Latin America, Japan and Korea will see growth continue in the 21st century.
Source: EMC World Cellular Database

Current mobile networks are feeling the pressure of exceeding their design specifications. Nobody expected 70% of the population to have and use mobile phones when the current mobile technologies were standardised some two decades ago. The new UMTS environment is designed not only for large numbers of users, but also for varying types of services on the network. New UMTS services are enabled with a QoS (Quality of Service) model for the terminal as defined in 3GPP[3] global standards. This model has several service classes ensuring that the radio connection is capable of supporting various types of applications:

- *conversational real time traffic*, such as multimedia conferencing
- *real time streaming traffic*, such as online audio/video reception
- *interactive traffic*, such as Internet browsing
- *background traffic*, such as downloading of mail

[3] 3GPP is the standardisation body for 3G WCDMA technical specifications.

Operators will be able to define the QoS level for each UMTS service depending on the price the customer is willing to pay. For services with a higher QoS, like video streaming, customers will be willing to pay more. Services that are not delay sensitive like e-mail can use the background traffic QoS class but will be charged at a lower rate than premium delay sensitive services.

Preferred device

The mobile phone is already the preferred voice device for hundreds of millions of users. Why? Because it is personal. It is the only device that is in our possession 24 h a day. It can contain all our important phone numbers, with names so they are easy to remember and find. It can be our diary, our notepad and now our access to the Internet and information that we need while moving around. Yet there is still one thing that many people fail to appreciate about the mobile phone. It is not just about voice or data or accessing people or content. It is how the mobile phone can reflect individual personalities, lifestyles and our moods The popularity, and operator profits of personalised ring tones are a clear indication of this. Here we are experiencing the first signs that consumers and business users in the near future will expect and demand unique and personalised products and services. The companies who recognise and act on this knowledge will be the undisputed leaders in their respective markets.

Now it is necessary for us to state the obvious. UMTS will be about services not technology. Even more than that, UMTS will be about management of our time and content. Technology such as WCDMA (Wideband Code Division Multiple Access) and IP (Internet Protocol) are only the enablers for the Mobile Internet. We all know that, but what does it mean? Basically if you do not get your act together and create what we call 'invoiceable' and personal services you will be a has-been in the Mobile Internet. The winners will be those companies that can create a multitude of user friendly services that people will pay for. Unlike the Internet community where 'free' is the byword for service, mobile users pay for their communication. They pay for voice since they value it. They are willing to pay for text messaging because it is seen as superior in so many situations where for example a fast, short answer is needed. Mobile subscribers pay for WAP (Wireless

Application Protocol) access if it provides good content and for many WAP is a valuable service that they are willing to pay for. It is relatively simple to extrapolate this to a situation where users are willing to pay for new value creating UMTS mobile services that are individually personalised.

This is where the arguments start. Or should we say the interactive discussions begin. Users will not, and in fact they can not pay indefinitely for more and more content. We all have a limit to the amount of disposable income we have for our personal communications. But this view will change. Lets consider how much of our allowance we spent on telecommunications when we were teenagers. If you are over 25 like we are, then the answer is zero. Maybe a few calls to the girlfriend or boyfriend from a payphone, but mostly our 'telecoms spend' when we were teenagers, were calls from our parents' phone at home.

Spending substitution

Teenagers today in Finland spend up to 90% of their allowance on their mobile phone bill. Over 50% of this bill can be SMS (Short Message Service) text messaging. This group of society are spending less on clothes, cinema and eating out. They are using text messaging for chatting, sending jokes, sharing simple pictures, providing information on what, where, who and how to their friends and even dating over SMS. The youth market do not leave voice mail messages since this is too time intensive. Its not instant. They send a text message. It is faster and more simple. We are now more likely to send work related text messages since it is less intrusive when people are in a meeting. Yes we admit that we are sending more and more text messages, but don't tell our kids.

We think the trend for UMTS is becoming clear. In the early 90's when GSM (Global System for Mobile communications) started it was the business user who was the target customer. In those days there was no youth market. No pre-paid. No text messaging. No mobile access to e-mail. Who even had reliable e-mail back then? Penetration rates for mobile phones of 30% of population were considered futuristic, for the dreamers, unthinkable. How we relearn and re-evaluate our opinions. What is certain is that any UMTS operators who ignore any market segment do so at their peril. We believe that all operators have to be ready to target multiple markets from the beginning and be prepared for the mass market take-up for any service from day one. It will be the mass market that generates the revenue growth. The mass market is where the greatest potential is. The mass market will be the early adopters. Kotler will have to re-write Marketing Management.

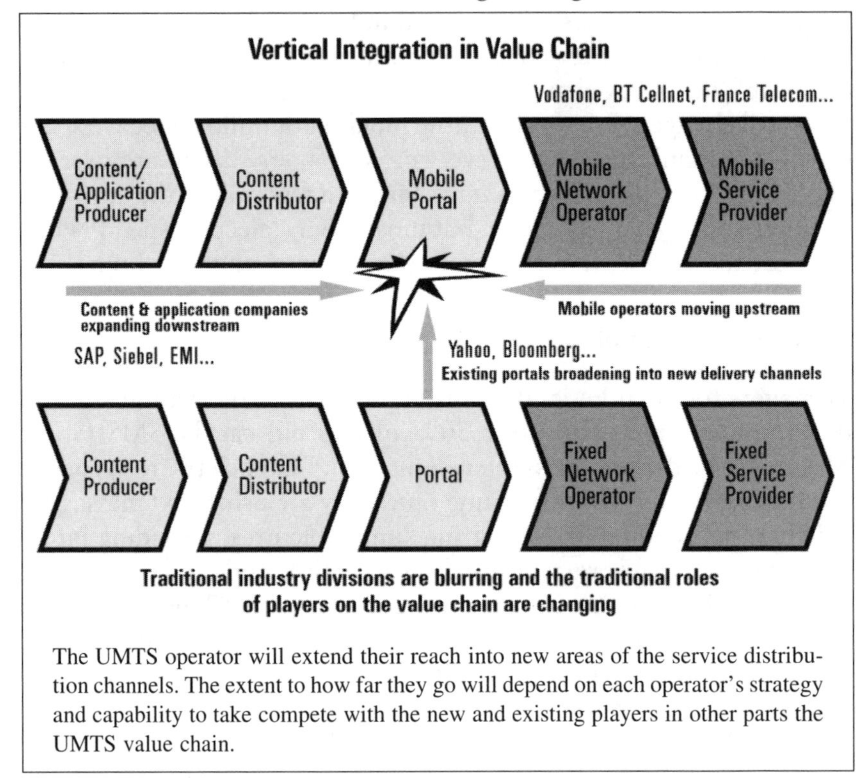

The UMTS operator will extend their reach into new areas of the service distribution channels. The extent to how far they go will depend on each operator's strategy and capability to take compete with the new and existing players in other parts the UMTS value chain.

1.2 Fixed internets, second generations, and UMTS

The services developed for UMTS networks will be products of the most complex, interconnected and intelligent machine man has created. It is at the heart of the convergence of fixed and mobile networks, voice and data, the existing fixed and emerging mobile Internets, and the convergence of digital content and wireless delivery. These various trends that relate to the overall convergence in telecommunications will have a great deal of impact on the UMTS environment. It is not the purpose of this book to go into depth regarding these networks. Chapter 14 briefly covers the technical side of the UMTS network but readers who want to learn more should refer to the book 'WCDMA for UMTS' by Holma and Toskala, also published by John Wiley & Sons.

How browsers changed the internet

From its birth in the 1960's the Internet looked and felt basically the same until the early 1990's. Techno-elitist researchers, mostly from America with a few West-Europeans, primarily using mainframe computers with Internet connection to different forms of person to person(s) communication and the exchange of files. Nobody had heard of the 'Worldwide Web' or WWW. All that changed when Mosaic was launched as the first WWW-browser and the Internet was never the same again.

Still in the early 1990s the Internet had its own decentralised information sharing system called Gopher. Many universities and organisations which had Internet connection in those days had their own Gopher homepages. It was university students who started to experiment with the Internet and became the first users of non-academic services, like checking the daily menu of their university cafeteria from Gopher just like they can do now on the WWW.

Even after seeing Mosaic and WWW, there were many devoted Gopher users who believed that the WWW was nothing more than a facelift of Gopher. It had a nice graphical interface that could display online pictures and it had a hypertext-structure which made page creation easier. The early thinking was that the WWW could ***never*** replace Gopher. Gopher already had a huge amount information and nobody would convert it to WWW-format. Gopher had logical hierarchical structures while the WWW was an incomprehensible mess of hypertext. Creating content for the WWW would be too difficult for typical end-users because it required a new mark-up-language while Gopher worked mostly with text-files. And besides, 'all users' knew how to use Gopher already.

How wrong those predictions were. Remember that these were near unanimous opinions by the best Internet experts and users of that time. As we move towards the world of mobile data services those lessons should be remembered. In only a few short months the WWW had more information than Gopher had built up in a decade. Its graphical interface and hypertext-structure provided excellent usability and end-users quickly learned to create new content. The Internet was transferred from mainframes to PCs. It became a mass-market service. It became commercial. The scope of its services widened. Today hardly anybody bothers to think of how a web page might seem to a mainframe user, but every content provider tests pages on the current WWW-browsers, Internet Explorer and Netscape.

How mobility will change the fixed internet

A similar transformation will happen in the UMTS future when content migrates from the fixed Internet to the Mobile Internet. It may seem like heresy to the hundreds of millions of users of personal computers, but already mobile phones outnumber personal computers by a factor of 3:1. Very soon most of the mobile phones will be Internet-enabled. The transformation is inevitable.

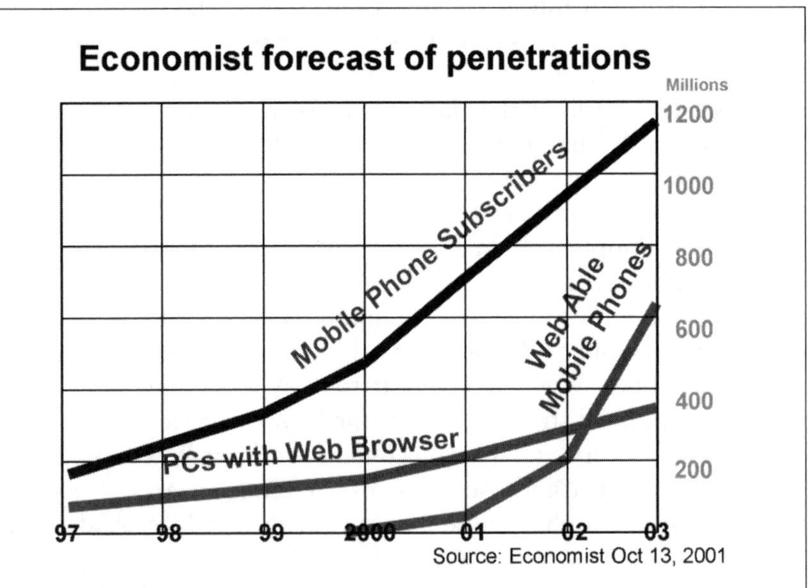

It is now widely recognised that the number of web enabled mobile phones will overtake the number of web browser PCs. In some countries like Japan and China, the first experience of browsing the Internet will be from a mobile phone.

The Internet will be accessed by a multitude of different devices; it will become more international; its business-logic will change and it will have new or at least enhanced services. The predominant Internet access device will change from the PC used today to the mobile phone in only a few short years. The content producers will write their primary content to be delivered by the most prevalent device – and that will be the UMTS mobile terminal. Most content will migrate from the fixed Internet to the mobile world faster than the transformation from Gopher to the WWW.

From client-servers to clients-profiler-servers

Currently most of the Internet users access the Internet only from one device: a PC (Personal Computer) either at work, in the home or at school. Some people use other means such as Internet cafés and libraries but this is still a small percentage of total usage. Several new technologies are being introduced to allow Internet access via other devices like digital TV over satellite. There are also small pocket size devices including PIMs (Personal Information Managers), PDAs (Personal Digital Assistants) some combined or integrated with mobile phones that allow Internet access.

In a few short years it will be common to use multiple devices to access and receive mobile content and browse information that has until now been primarily available via the fixed Internet. Many people will of course use their PC or similar devices at work as their preferred device to access the Internet, then, on the way home their cars will connect their navigation and information systems to the Internet. At home people will be consuming content via the digital television which will download various types of content from the Internet like pay-per-view movies. While watching TV many people will have an Internet access device such as an Internet browsing tablets to enhance and supplement their TV viewing, for example while watching a sports game, to check on the status of scores in other games or to brows other Internet sites.

Throughout the day that same person will use the UMTS device to access the mobile content for various services of convenience. It is important to recognise that in the near future people will access content from a multiple of devices and in a number of different ways. We will no longer have only one access method. This is very similar to the way the use of radio developed. Early on, families had only one radio, and the whole family would gather around it to listen to specific programmes. As families started to gain more radios per household, they also started to tune in and out of favourite channels from different radios during the day, from the wake-up clock-radio in the bedroom, to the kitchen radio, to the car radio, etc. The variety and usage increased so that definite identifiable segments emerged with differing listening patterns.

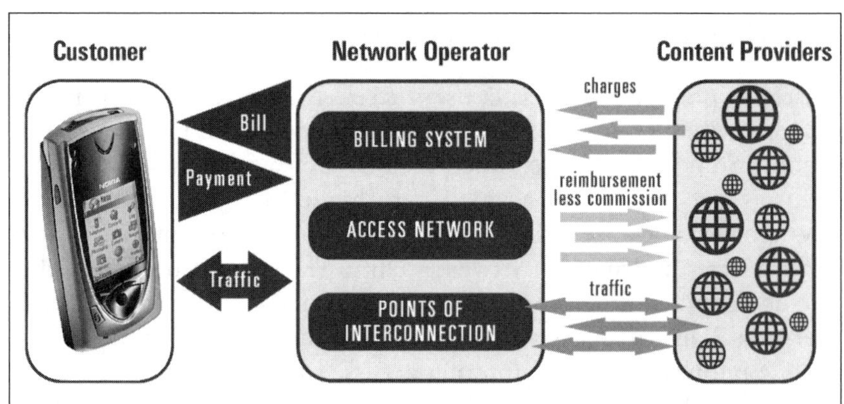

The value chain becomes more complex and the number of "links" increase in the UMTS business. The UMTS operator now has the chance to create a one stop shop for all connections and access to content while mobile.

Currently end-user information is shared between PC client and the Internet server. A simple example is that the browser's bookmarks reside on the PC while the e-mail inbox resides at the e-mail server. As people will access the Internet from multiple devices, more information must be stored in the network otherwise the current device may not have all the information needed to carry out the transaction. Amongst other things this means a radical re-thinking about Internet 'cookies'. Cookies are small files that web-sites send to user computers to recognise the users and let the service be personalised. When users return to the same site these files are sent back and the site knows immediately any relevant information about the visitor. This is why many services have you sign up for the first time with personal information, and then when you log onto the same site later, you are not asked those questions. But whenever the you access the same site from *another* computer the web-site cannot recognise you and the sign-up routines are called up, and a new cookie is created at that computer to identify you.

User profiler as a service

Multiple devices means the end for the current predominant internet structure of client-server – or rather one client – many web-servers architecture. As we get ever more devices that we use to access the same content, we will not want each of them to ask us to identify ourselves. We want our profile to be managed centrally in one place, and then, when we access the sports scores from the office PC or from the car or from the home digital TV or our mobile phone, we want all of them to know what sports we follow. This creates a need and opportunity for profiling as a service.

The future architecture may be something like many clients – one profiler – many web-servers. This presents an opportunity for the UMTS operator or the portal service to run a profiling server that stores end-user profiles. This system would replace the current cookie structure by keeping track of users, their preferences and past transactions across multiple access devices even under multi-session simultaneous use. This profiler should be able to authenticate the end-user no matter what device they are using and send the user profile to the web-server. This information could logically relate to mCommerce transactions, location information, and of course billing information.

The profiler service would predominantly be offered to content providers and portal services, etc., as a way for them to build customer service and personalisation serving the needs of the partner who wants to build a successful service for UMTS. The profiler service would be a key differentiator and a particular strength of the UMTS operator. In a very real sense it would be a gatekeeper function and the operator can take advantage of this position.

1.3 Recent service trends

At the writing of this book, no market existed with UMTS services in full use. Early service launches intended for Japan, the Isle of Man, Spain etc., had been delayed, or only limited trial networks been deployed. Thus no real data on real UMTS service use existed for us to use. We have examined a few closely related technologies and geographic regions to give some insight into what can be expected.

American lessons

The fixed Internet community has been heavily focused on America. Personal computers took-off more rapidly in the US since fixed line telecommunication was first released to free competition there. With computer networking evolution the US went directly to the Internet unlike France for example which tried to retain its national Minitel system. Today France is still way behind in the number of PC's connected to the Internet when compared to other similar countries in Europe. Although Minitel was a totally successful service, well liked by all users, operators in France have recognised that it will never compete in the long term with the Internet. In the early adoption of broadband solutions to the fixed Internet, again the US is leading. Clearly the Americans lead in the adoption of the fixed Internet, and many valuable lessons can be gained from following the fixed Internet experience in the US.

European lessons

When looking at mobile telephony the roles and development experiences are reversed. What the US did right in the fixed Internet it did wrong in mobile networks and what Europe did wrong in the fixed Internet it did right in the development and growth of mobile phones, especially in the move to a single digital standard. Europe, unlike America adopted a common digital mobile standard; GSM, allowing easy roaming of services and pricing of services to generate traffic. Many European countries lead the US by a wide margin in mobile phone penetration, mobile phone traffic, text messaging and mobile phone revenues.

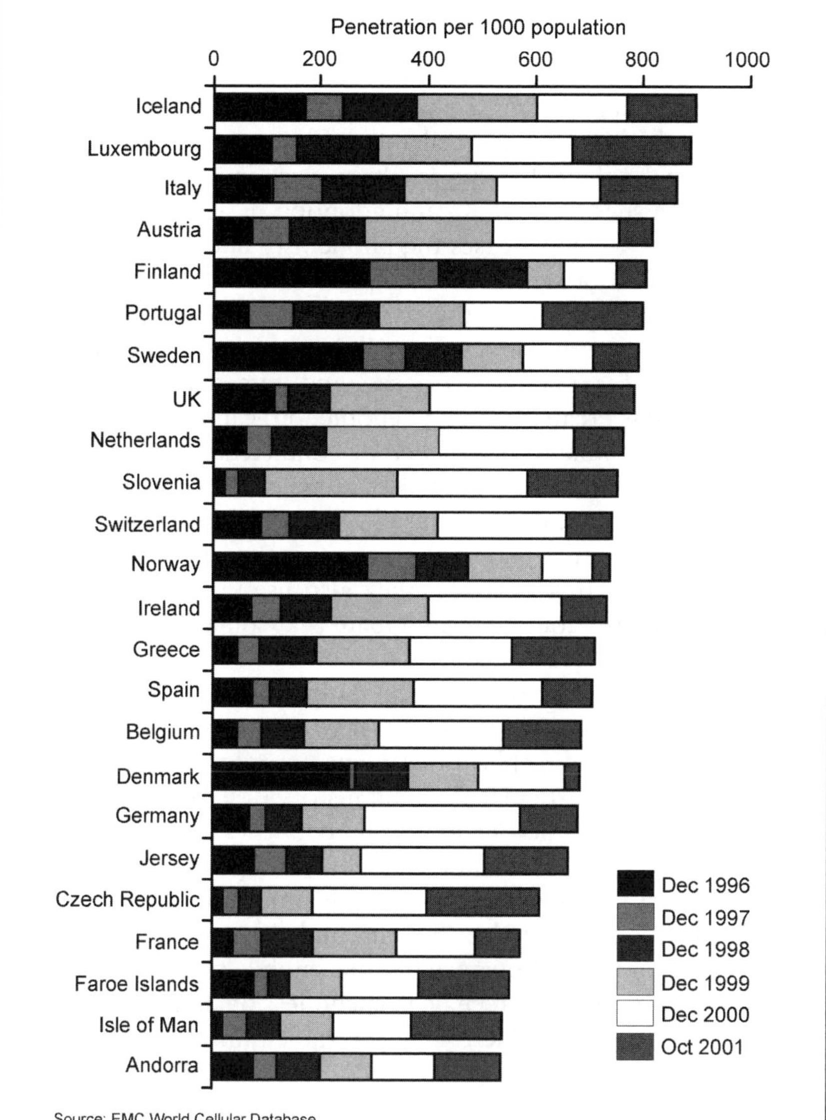

Many developed European countries are reaching their maximum mobile population penetration. Growth in the future will then rely more on new data services. Source: European Mobile Communications Report, published by EMC, November 2001.

A key differentiator in the adoption and use of mobile services in Europe is the invoicing of calls. In the US, cellular phone users have to pay to receive calls. Therefore they tend to keep their phones in the off position and cellular phone users usually have a pager (beeper) to let them know who is wanting to call and the phone user will turn on the phone and make the call. This results in very inconvenient use and strongly diminishes the take up and adoption of wireless services.

In Europe for the most cases the phone user does not have to pay for receiving calls (a significant exception is when roaming abroad where the called party pays the roaming part of the call) with the result that phone users keep their mobile phones switched on at all times, allowing for much easier access and immediate, spontaneous contact. This has created a new phenomenon, called 'reachability' – the state of being able to be reached or contacted via mobile phone. Reachability has been used to explain some of the dramatic unanticipated growth in mobile phone use in Europe, not seen in America. People not only use the phone to place *outgoing* calls which the user can anticipate wanting to make; they also use the phone *to receive calls* which the user could not anticipate that someone else wants to complete.

For anybody wanting to study the mobile cellular network environment, Europe is the place to be. There are numerous fascinating services, bundles and applications being introduced around voice, SMS and WAP. In particular the long history of mobile communication competition in the Scandinavian countries make places like Finland, Sweden and Denmark very fascinating areas of study for emerging trends in mobile telephony. By example, the teenage market is almost 100% penetrated in Finland and this group spends up to 90% of their monthly allowance on their mobile phone bill. The flip side of this is that Finnish teenagers spend less on clothes, eating out and going to the movies. Here is a clear indication that behaviour in this social group is changing.

Asian lessons

The Chinese and Japanese have been behind in Internet penetration partly because their written characters have been difficult to implement to standard computers using traditional keyboards.

Although current WWW browsers and servers support Asian character sets, the WWW hasn't reached the same popularity in these countries as it has in the West. Mobile phones have had Chinese and Japanese displays for a long time and when pen-based user interfaces become more common in mobile phones there is no reason why the Far-East would not rush to the fixed Internet. The sheer size of the Chinese market, for example, is such a potential, that a successful new service there can have a dramatic impact on the whole industry.

There is constant talk and discussion currently on the overwhelming success of the i-mode service from NTT (Nippon Telephone and Telegraph) DoCoMo in Japan. The service has had phenomenal adoption, penetration, usage, growth, and profit numbers. In two years the service accumulated nearly 20 million subscribers, the majority of who are accessing Internet based services for the first time. There are many reasons why the Japanese market may have been particularly suitable for the i-mode service and perhaps not all aspects of the phenomena will replicate globally, but clear early lessons can be learned.

The service usage costs have been kept below the pain threshold, so that with most basic services the costs to users have been perceived to be low enough so they do not to stop and consider whether they want to consume the service or not; they just do. Many services have been built to create a strong attraction, almost addictive or viral in nature in some cases and the ability to communicate with one's friends, the community aspect has allowed easy word-of-mouth to spread interest in the new services.

One should not forget the other mobile operators in Japan, where J-Phone and KDDI have achieved remarkable subscriber growth and mobile service usage numbers. These would be headline news were it not for the even greater success of DoCoMo and i-mode. From a practical point of view, the three major mobile operators in Japan have competed against each other in Mobile Internet services in that market. Readers of this book should follow very carefully what is happening in Japan in order to gain some insight into the potential of UMTS services in the near future.

In other areas of Asia SMS has been hugely successful, especially in countries like the Philippines and Singapore. Initially SMS

1 Vignettes from a 3G Future

Picture Postcards

I have arrived at the Eiffel Tower in Paris. I place my UMTS phone on a table and click on the self-timer for the camera. I take a picture of myself with the Eiffel Tower in the background. I look at the picture on the colour screen of the UMTS phone and decide I like the picture. I write a greeting to go with the image "Hi, Paris, me and the Tower." I select from the UMTS phone's phonebook my sister's name and press send to send the picture postcard to my sister. Then I select my cousin, my nephews, my parents, etc. and end up sending about a dozen postcards. All without looking for a card, stamps, or a mail box. Without the trouble of bringing the mailing address, or writing separate greetings. And every picture postcard would be similar in cost to a regular postcard and the postage stamp.

A telecoms operator will be able to offer picture postcard delivery as a relatively low capacity network cost, as the image will be still, requring little image data, and since the image does not have to be delivered in real time nor immediately. Operators can deliver the picture postcard in "background data traffic class" mode and price such data transmissions as relatively inexpensive data communication.

was given away free as an incentive for customers to join the GSM service. SMS usage though quickly grew to the point that operators were forced to start limited charging for SMS usage as the impact on the network was so large. Some of the Asian operators are delivering over 100 million SMSs per day and the percentage of revenue from SMS is approaching double that of European operators. Free SMS can only account for part of this growth. It has also been driven by the need to stay in contact, to have simple communication methods with family and friends. Many of the new SMS based services like wireless pets and SMS payment and banking solutions originate from Asian operators keen to exploit existing technology and boost usage and profits.

1.4 Money lessons from the fixed internet

Simple fact number one: to make money out of a service based on the access to mobile or fixed content (Internet), someone has to be willing to pay. Simple fact number two: in the long run nobody pays more than the value of a product or service. Simple fact number three: most items of information and entertainment consumed currently on the fixed (and free) Internet have only slight value measured in pennies, not dollars.

The real value of individual information and entertainment on the fixed Internet is diluted by the sheer abundance of redundant sources. You can get sports scores from the newspapers for small change, or for a relatively small fee on TV. You can get the sports scores from the fixed Internet from hundreds of sites, almost all of which are currently free. The usefulness of getting the information a little bit earlier rather than waiting for the TV news and of being able to scroll through relevant stories, has only small marginal appeal to even the most avid sports fan. So the value of a single sports score update on the Internet might only be a couple of cents.

But most billing systems have individual consumer billing costs running at anything from 5 to 20 dollars per delivered invoice, or often much more. Credit card companies have minimum charge limits in many countries which tend to be of the magnitude of 7–10 dollars.

So the big challenge has been how to make money in the fixed Internet if the value of the product is less than costs of charging for it. Currently many sites providing only marginally valuable services have counted on advertising revenue. For example, Yahoo's services, stock information etc. are free to users and advertisers are almost the only revenue source for those Yahoo! Services. The other logic has been to sell items that are so valuable, that charging costs don't become a barrier. Amazon's web-bookstore has been the most popular example of this model.

Magic of micro-payments

Handling very small amounts of money, 'micro-payments' are discussed in more detail in the next chapter on UMTS service attributes. Micro-payments with a value of less than 1 dollar and often mere pennies or even fractions of a penny, will provide new possibilities to charge for products and services that could not be charged for in the fixed Internet.

These costs are small enough to be priced well below the usage thresholds for any given service and the billing systems of the UMTS operators provides a unique competitive advantage to build micro-payment options to bill for services of only small utility and slight value. This is the first time that most fixed Internet content providers have a real opportunity to build a system to get revenue for the *value* of their content. For any struggling Internet content provider, this is the opportunity they have been waiting for. And as the tiny trickles of revenue streams are multiplied by the hundreds of millions of potential visitors from the mobile world, the content providers will all be eager to develop *better* content for these customers. All they have to do is make sure that their content is mobile aware and suitable for delivery to mobile terminals.

The tiny revenue streams from the small transactions, when multiplied by the millions of users on mobile networks, result in very large income streams to the content providers who are now struggling to get anything from the fixed Internet. This is the magic of small payments and large populations; and the business logic of micro-payments. Early on the micro-payment model will only exist in the

UMTS mobile world. In the future it is very likely that common micro-payment systems and standards will also emerge on the fixed Internet as both mediums merge into one ubiquitous service offering.

Both fixed internet and mobile internet are needed

The fixed Internet and personal computers with big screens and broadband access have excellent access to a world of information, but the user is tied to the personal computer. The UMTS mobile world can be taken anywhere, but the content is less easy to digest due to the small screen and due to the greater costs relating to airtime. Which is better? Neither. Both will exist. Let us use an analogy from public libraries. Reading books and magazines is free in the public libraries. This is not a new invention, it has been so for hundreds of years. Yet still there is a flourishing market for the very same books and magazines which are bought for the home.

Both have found a place in the 'information' marketplace. People buy books and magazines, and they also borrow books and magazines from the library. Perhaps we prefer one or the other based on our standard of living, convenience, stage in life, even particular subject matter. Occasionally one might need a specific book which might be out of print, or maybe there is the need to do research into many books that one might not want to buy, or there is the desire to find a back-issue of a magazine. On all these occasions it is possible to get the information or service from the library. But it is inconvenient to go to the library to read, or to borrow free books which then have to be returned. We have less time and there is a 'hassle' factor that we do not like. As an alternative we buy books and magazines. We don't buy everything, but we find definite value in having specific books and magazines at home *even though they are available at the library for free*. There is a place for the book store and there is a place for the public library.

Similarly one can access the fixed Internet and its content, for free (or at the low cost of Internet access, depending on country). But it is inconvenient and tied to the location of the Internet access devices, mainly personal computers at home or at the office. Now it is possible to also access information on the mobile terminal. This will be much more convenient in the future especially with UMTS service and is not tied to any one location. But the added utility of the access convenience would come at an access price.

1.5 The end of the beginning

The potential for UMTS services is enormous. The growth will be mind-boggling. New players and existing mobile operators will be scrambling to take a profitable piece of a rapidly growing pie. What are those services, how will they be marketed, who will buy them, and which players will emerge at the end? This book will look at the early thinking into those questions.

As we said in the beginning, UMTS is not about technology, it is about services. Those services will need to be desirable. Those services need to be personalised. Those services need to move with us as we want them. Those services need to be invoiceable. Those services need to be profitable. The UMTS service environment is very different from any of the existing service environments from any of the telecoms, IT and content worlds that are converging into the future. UMTS may seem bewildering, remarkably complex, and at times downright counterintuitive. Yet one of the true masters of capitalising on technological breakthroughs, Henry Ford, gives us his insight, and partly even explains the structure of our book, when he said: "Nothing is particularly hard if you divide it into small jobs."

2

'Basic research is what I am doing when I don't know what I am doing.'

Wernher von Braun

Attributes of Services for UMTS:

What makes for desirable services

Tomi T Ahonen and *Joe Barrett*

There are probably few operators who are now looking for the 'Killer Application'. That one service that they can claim justifies the money that will need to be invested in licenses and networks over the coming years. Most people now agree that there is not one service that will make the market. Some claim it will be a number of small killer applications, or that it will be personalisation of services that are tailored to individual's needs. We like to think that successful UMTS (Universal Mobile Telecommunications System) operators will create a 'Killer Environment', a place where it is easy to create new services and experiment with unusual applications, to learn more about customers and to give users the wide range of choices that will make them use their UMTS mobile device for everything from paying for bus, train and air tickets to downloading music video clips and short movies.

UMTS operators will need to create thousands of different services, each with its own combination of target market, availability, pricing, usage pattern and predicted traffic load. These services need to planned in such a way that the UMTS network will be able to support the service and deliver the QoS (Quality of Service) and user experience intended by the Marketing and Service creation managers for each individual service. Operators are moving towards a more service creation orientated organisation with less of a main focus on the networks. Some may even split their business into separate networks and service delivery companies so care should be taken that the UMTS service matches the network's ability to deliver the planned user experience. UMTS services will have to be designed to match the network technical capabilities in the mobile core and radio network with the support of new evolving back office solutions like billing, security and content provisioning platforms.

Each service will probably need some kind of partnership for content creation and update, and many will need custom development or significant modification from the existing platforms, media and solutions. All of the services will need updates, maintenance and must respond to competitive pressures from services offered by other UMTS operators, and of services delivered by other media and technologies.

2.1 Competition from old economy and beyond

For example, news services provided via UMTS will become more competitive as all UMTS operators will launch and upgrade their own news services. But TV, radio and newspapers will all feel the competition, and will be forced to innovate, introduce new and better services on their existing platforms, and to cut prices. Other telecoms media, such as GSM/WAP (Global System for Mobile Communications/Wireless Application Protocol) and the fixed internet will bring added competition. Wholly new technologies and solutions will emerge, including hot spots and digital satellite TV.

So what makes for the most successful services? Some things are not likely to become very successful on UMTS. One could build a

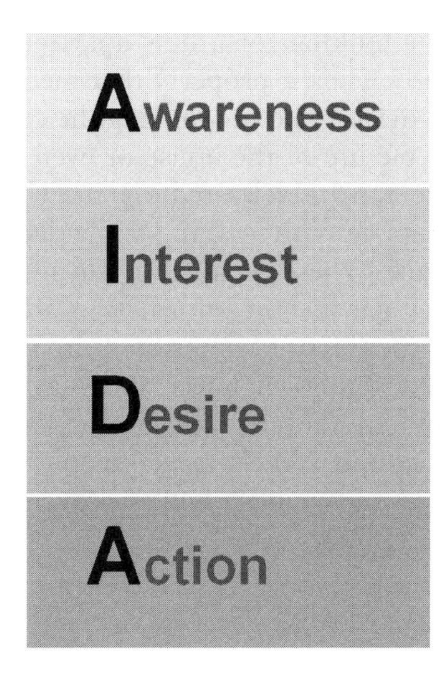

Figure 2.1 AIDA buying process.

value chain and partnership to purchase luxury automobiles or houses or sailing boats via the mobile terminal. The purchases are not typically done rapidly in a short amount of time, and the small screen would not make for an excellent presentation media, and mostly this kinds of luxury purchases have a lot of 'touch-and-feel' aspects that are not easily replicated on a mobile phone terminal. What the mobile phone can do is participate in part of the buying cycle for higher value goods or services that would not ideally be purchased without this touch and feel Figure 2.1.

We can consider AIDA (Awareness, Interest, Desire, Action) as one view of the buying cycle as shown in Figure 2.1. This starts with **Awareness** of the offer or competitive advantage in the market, **Interest** in this advantage, product or service, a **Desire** to find out more information on the offer and finally if the perceived value is equal to or higher than the cost, **Action** that leads to a sale. The mobile operator can play a part in the buying process for larger purchases mainly in the Awareness, Interest and Desire part of the cycle and less of a role in the customer's final decision or Action. As an example a couple looking for a new apartment or house would like to know instantly when a property that meets their criteria and price limit is on the market. Before making the call to the agent they may prefer to see a picture of the house or even a short sales video. This kind of service can be delivered via the UMTS network to a UMTS terminal equipped with imaging and video capabilities at any time or to any place. There is a clear advantage to the mobile subscriber since they get instant information about the availability of apartments that they may be interested in. The agent is supplementing existing marketing channels and may even be replacing them with this kind of mobile service that can be at a far less cost. The agent is also more likely to get people viewing the properties that they are seriously interested purchasing resulting in less wasted time, again resulting in less costs.

Colour screens and imaging phones will integrated camera and video capabilities will help create the early UMTS market. This will start before UMTS with terminals like the Nokia 7650 imaging terminal that also supports GPRS and HSCSD.

This kind of service is unlikely to play any major or minor role in the purchase of a new car but it can play a very large part in the buying process of second hand cars, even quite expensive models. This does not only work with high value items. Pictures of the latest fashions delivered to the mobile phone can create Awareness, Interest and Desire to go to the store to try on the clothes, increasing the likelihood of a sale to the retailer. This kind of service can be augmented with the delivery of discount coupons. Once mobile operators appreciate the role that a mobile service can play in the buying process the better they will appreciate the kind of sales supporting services that will add value to their customers.

Services should be personalised

The personal attribute for a service makes it feel to the user that it is genuinely unique to that person. If one is a golfer, then services about golf, or advertisements, discounts, maps to golf courses, golfing weather, etc. can be of use. But if the person does not care about tennis, then similar news, advertising and content about tennis is irrelevant. It is very important to build services for the particular interests of the user, and to try to focus the user's interests very precisely.

Many degrees of this personalisation can occur. For example if the person likes fashion, then probably fashion news and advertising is welcome. If the person likes Italian shoes and handbags, and the fashion content is filtered so that not even French shoes or Italian scarves are included, the more personal it becomes and thus more relevant.

What is important to recognise at the start of the data services revolution we are entering is that although everyone has their own specific and individual needs, in effect their own personal profile, the capabilities of the network and the operator's organisation and processes is evolving. Yes we are approaching the position that we can have a market segment of one, but that is not where the operators nor the content providers are starting from. Personalisation will start with new segments around lifestyles such as sports, business, music and games and evolve into more targeted and personalised offers. This will be made easier as the networks become more technologically sophisticated, the content delivery platforms become more flexible and the mobile terminals have new features like Java.

2.2 *Micro-payments*

By definition a network operator is positioned in between end-users and the content consumed by them. The operator may produce and host some or all of the content, but the network it operates will still serve as the distribution channel. This network has functionality for keeping track of numerous events, as the system periodically aggregates the usage data of each individual customer in order to produce the monthly user invoice. The aggregation of call data is a complex procedure, as a single event is rich with information such as call duration, called number and tariff information. In the UMTS network, and even in GSM with the introduction of GPRS (General Packet Radio System) packet data, this aggregation becomes even more complex. Mobile operators have had to build their invoicing processes around micro-payments and are the only industry that have

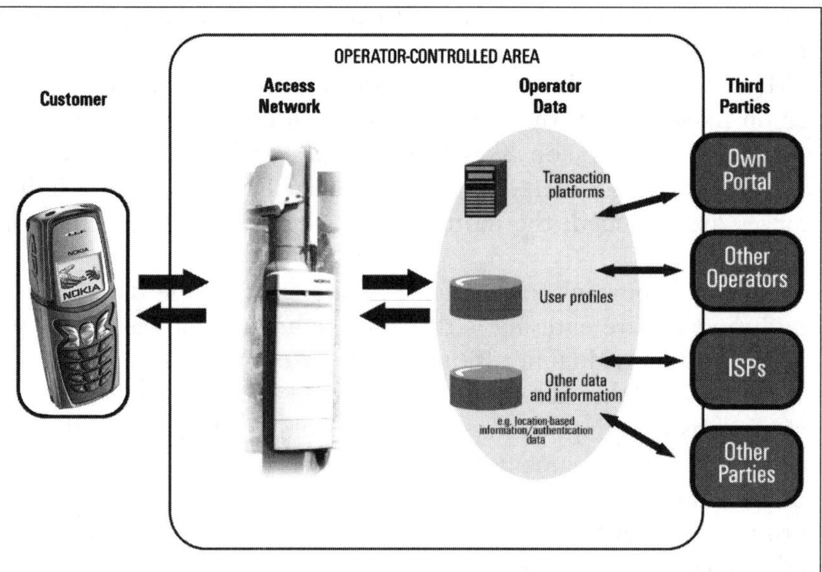

The UMTS operator clearly has control over its own domain - the network and back office systems as well as control over the customer's access and transactions. Third parties are necessary partners in various areas of the business to encourage growth and diversity of services. The fundamental control though is that of the transaction platforms user profiles as well as historical data and information of each users activities.

had to do this. This effectively places them in a powerful position in the future value chain where the majority of payments are made using the mobile phone. The electronic wallet of the future is not a smart card or any other separate item in your pocket, it is integrated into the UMTS terminal.

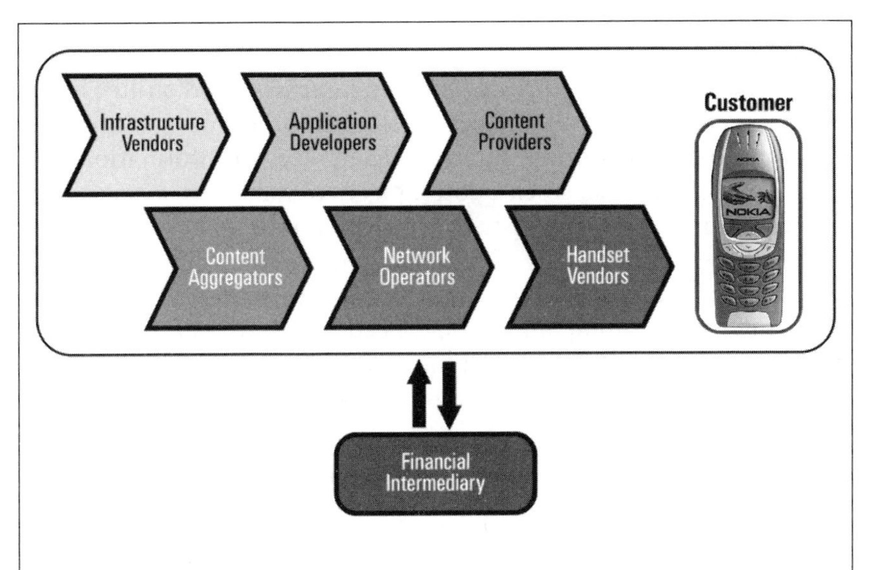

UMTS operators will work with Financial Intermediaries to provide payment and banking solutions. Customers are unlikely to move to a UMTS operator bank en mass since many of the banks are entrenched in their markets. The main component that the UMTS operator brings to the negotiating table is micro-payment capability.

Billing and micro-payments

The major cost associated with selling any service is the cost of billing for it. For some service providers the feasibility of introducing some new service may be ruined because billing for the service to be sold would cost more than what could be charged for it. Billing systems are typically large installations that are able to process very complex data inflows. Consider a typical telephone call: the billing system needs to be able to take into account the time of day, day of week, calling number, called number, call duration, possible discount campaigns, possible individual call contracts and various other information in order to arrive to a chargeable sum for that call.

All of this needs to be done in real-time for the vast number of calls that are taking place in the network simultaneously. At the end of the invoicing period, the billing system needs to compile all of this data for each individual subscriber. The cost of installing and running such a system is huge, and it needs to be recovered somehow, typically in charges from end-users.

The cost of billing may form an obstacle for introducing services in many types of environment. Consider an Internet-based on-line information service where the pieces of information sold are tiny and available to all users in the Internet. The pieces of information sold might be, for example, stock quotes, pieces of news in a specific area, or even menus of specific restaurants. It is very hard for this type of service to be commercially viable over the Internet unless there are thousands of individuals using the service an immense amount of time. Information companies would need to install their own billing systems to charge users over the Internet for content. This is unrealistic and again exposes the failing in the Internet business model.

The UMTS operator has the required billing system

In the Mobile Services World micro-payments are viable since the operator has the invoicing processes and systems in place. A micro-payment is a transaction charge that is too tiny to justify billing for each transaction separately so these can be aggregated by the Charging and Rating Centre possibly into types of micro-payment. Most users will not want to see the details of the purchase of one cola or one pack of gum or details of all the items under 1 dollar or even 5 dollars. However if they do and there is always someone, then this itemized list would come at an extra cost. The mobile operator is then in the ideal position to provide a service that no other industry can. One invoice for the majority of purchases. No need for cash anymore.

This can also lead to a more sophisticated accounting service for customers. The operator is in possession of all the purchase information including store and even the category of the purchase due to it being scanned in by bar code. By allocating accounting codes to each purchase it is possible to create a monthly spending summary so that customers can see where and when they have spent their money. Mobile payment solutions are already working today, are in some cases fully commercialised with hundreds of thousands of customers and are growing in sophistication. It is not too far fetched to imagine a time when the home accounting systems are as common place as in all businesses. The secret is to make the service so simple for the customer that it is worth the cost.

Do we want all this as users? Too right we do. When Joe moved to Finland in 1997 he found Internet banking the every day practice. It was a revelation. Not only that he could go into any bar and order a beer with his bank card he could get cash if needed or pay for a taxi with his bank card. Tomi had the opposite effect when he moved from Finland to the US in 1983. He was astounded to find that unlike Finland, where banking machines were commonplace and pay-checks were sent electronically automatically to the bank, in America most pay was paid by check, which needed to be deposited in a bank, and most payments were done by check.

Today we use bank and credit cards more and more so we don't need cash all the time and we tend not to carry it at all. The times when we do need a small amount of change for a micro-payment then becomes a frustrating trial of finding the closest ATM (Automatic Teller Machine). How much more simple to take out of your pocket your cash enabled UMTS device and pay with the press of a few buttons.

Are the banks likely to see this move to a mobile enabled cashless society as a threat? Possibly. But the banks offer far more services than just handling your bank account. They are more diverse in their financial services offering and mobile operators are unlikely to be able to challenge that position in the short term. However it is unwise to dismiss any possible future. The fact that the traditional banks have over 100 years of history and a strong position in the market can be a weakness. We only have to look at the FTSE 100 in the UK to see that long traditional histories do not count in the new Internet enabled world. Old stalwart companies like Rolls-Royce and Hanson have been replaced by younger more electronic and IT sassy companies that few people today have heard of. Vodafone did not even exist in 1986[1] but has accounted at one time for over 10% of the value of the FTSE 100.

[1] Vodafone was spun off from Racal Electronics in 1987.

So, the micro-payment business proposition seems very simple yet may be complicated to set up. However, it makes a lot of sense for all the parties involved. The network operator will leverage its position between content and the end-user and leverage its investment in existing systems. For the content provider, a micro-payment arrangement will make the roll-out of otherwise commercially unviable services possible.

2.3 *Further attributes*

The UMTS operator can take advantage of several other attributes of controlling the network and combining components and abilities. None of these are always necessary for a service, but many of these can be used in conjunction with the personalisation and micro-payments abilities to further enhance the utility of services.

Location-based services

An important feature of many applications operators and vendors are exploring will be based on location-based service delivery since this is one of the operator's key competitive advantages over the Internet. These services can include location-aware information, games and community services. The main driver for mobile services is making the data service relevant to the user while they are mobile.

A great deal of content delivery can be based on geographic location, local promotions, maps, weather, news. Operators can earn promotional revenue because they know the user's location[2], personal profile information or segmented channel. What is important here is that users will not accept promotions that are not screened and profiled for their relevance. We are all frustrated by the amount of junk mail that we get every day and we will not tolerate this happening in the mobile world. Push promotions will have to be either subscription based so that the user can

[2] According to Cell Id and RTT (Round Trip Time), IP-DL (Idle Period-DownLink), Assisted GPS.

indicate the information of relevance and interest, or they will result in a reduction in the users monthly bill or will be tied to a free service. There have been many proposals that users will get a promotion as they pass the cinema or shop but users will not accept a bombardment of promotions as they take a short trip down the high street. In this case the 'off' button becomes an excellent screening solution. Not the best option for creating demand and increased usage.

Multitasking

Any application or service which allows for multitasking can result in a valuable service in the UMTS environment. Any places, occasions and situations where time is 'wasted' could be recouped. These include all instances of waiting, queuing or temporary delays. Also any cases where some monotonous and low level participation is needed, such as sitting on a bus or waiting in an airport is ample opportunity for multitasking type solutions. It is at these times that subscribers will take the opportunity to have a little fun, say playing a mobile game or reviewing the picture messages sent from friends. This may also be the opportunity to review the news stories sent during that day or check the situation of your stock portfolio. As people gain new mobile ways to access and send information they will take the opportunities at different times of the day to do a variety of service transactions. The typical mobile subscribers day will then be made up of many mobile experiences some lasting a few seconds and others lasting minutes.

Text-to-voice

One of the primary solutions in multitasking is voice. If e-mails or news are read by voice applications and 'spoken' to the receiver, then e-mails can be received while doing other things, such as cooking, driving a car or gardening. Many forms of data might be able to be converted to voice translation and delivered in times when reading is not possible or even desirable.

2 Vignettes from a 3G Future

How Long will This Rain Last?

I walk out of a shop onto the street and it starts to rain. I look up at the sky and return to the shop for cover from the rain. I have to decide should I take a taxi, or stop by at a coffee shop, or perhaps buy a cheap umbrella. To help with that choice I take my UMTS phone and connect with the local services for local weather. I ask the "How Long will This Rain Last?" service that tells me that this very rain shower raining upon me will last for 7 minutes. I know I will be charged about 25 cents for the info and am happy I checked. I go the nearby coffee shop and have a cup of coffee while the rain finishes. And it never ceases to amaze me how incredibly accurate these personal weather forecasts have become.

The personal current weather info service is actually a combination of several services and functionalities. The UMTS network has the ability to locate the exact position of the UMTS phone within several meters. Then a special real-time weather conditions service is accessed which uses weather radar like those used at airports. The local weather radar will determine which rain cloud is currently above the UMTS phone. Then by plotting the speed (velocity) of the weather pattern, and the direction, the weather service can determine very accurately short-term weather changes, such as how long it takes for the specific cloud to go past the UMTS phone (and person) below. This kind of real-time weather radar info is much more expensive to collect than standard meteorological data, but if small "micropayment" sums can be collected to pay for the service, the costs of the weather radar info are easily covered and the service can be made very profitable.

Multi-session

The multi-session abilities of UMTS technology will allow services that are used simultaneously. Most typical will relate to combining the telephone call and viewing something at the same time. A typical example is looking at the movie listings and talking on the phone with a friend who is also looking at the same listings. It is even possible to view a movie trailer with the friend and discuss it and decide which movie to go to see. As all of the services are digital and can be used in multi-session mode, then any combination of using them is possible. Reading e-mail and carrying on a conversation. Downloading a file from the office while paying bills on m-banking.

The technology behind this multi-media type of call is SIP (Session Initiated Protocol). SIP enables the signalling of each service to be separate. This means that two people can start with a simple voice call, add video if they want to see each other or show visually some item, drop the video and still maintain the voice call, share a picture or map and even use a shared white board. Each part of the multi-media call, voice, image, video or data transaction can be charged separately so that as the quality or value of the multimedia call is increased the cost increases and then falls as the multimedia 'value' is reduced. Since the multimedia call can be one or a combination of the above, voice does not have to be included with video.

Timeliness

Timeliness has several varieties. The first way to look at timeliness is that of immediate availability. For example we can all borrow a how-to book from a library for free, but there is considerable utility to having it on hand when you happen to need it. In the same way there is a great deal of information on the Internet and other sources, but the utility of a service is improved if you have access to the information when you really want it. Typical examples would be price comparisons when you actually go visit a store, etc.

These services can also have a variation on timeliness in the form of immediacy. Immediacy means that when the content changes, you get the notification immediately. In most cases immediacy is not relevant. For example the TV listings for the following week are usually collated about a week in advance. It is not particularly relevant to know exactly when the new week's listings are published, rather the info is collected often on the day of the programme. But if one is involved in politics, and a rival party has published a press release, that is immediately relevant. It is likely to have great impact and the sooner you know, the better. In the business world, if the salesperson knows within a few seconds that the price list has changed or if the competition have announced a new product or have some delivery difficulties then the sooner this information is known the better equipped they are to close the deal.

Yet another variety of timeliness is real time. Real time means that the information which is seen, is actually the latest available information. In many services and info content, the information may be delivered with a delivery lag. With fixed Internet applications, several layers of buffers exist where content may wait for hours, even days. If for example a sports service provides news on a 30 min interval, it is not very useful for following live sports when they happen. Equally a stock price ticker brings considerable value if it is in real time. Mobile stock price services are currently available but individual stock prices are not normally delivered – pushed to the mobile terminal in real time.

Attributes and how they are delivered

This brings us to an important point of services, their attributes and how they are delivered. Technology exists today to create mobile data push services that can start to create the type of usage patterns that will drive service evolution towards UMTS. Some companies already push content to their employees about stock levels, pricing information, competitor information, and company stock price. It is even possible to get the contact details of any employee in your company within a few seconds. These services uses the SMS (Short Message Service) to either push the messages to individuals who subscribe to the content that is pushed in the form of a message or in the case of the mobile phone book the sending of an SMS with the name of the person you need the phone number for. Some mobile operators have introduced this SMS mobile phone book services so if you know the name you can find out anyone's mobile phone number. What is surprising is that more operators have not used SMS more widely. It is almost as if the technology is too simple to be considered as a delivery solution for more than just person to person messages.

2.4 *Service creation aides*

Several aides exist to assist in defining and creating services. PAIR (Personal Available Immediate Real Time), MAGIC (Mobile Anytime Globally Integrated Customised), the 4 P's (Portable Personal Processor Proactive) and 0-1-2-3 (0 manuals, 1 button internet, 2 s maximum delay, 3 keystroke maximum) are among the best known, and each offers some good insight into how services can be created in the mobile services space.

PAIR

The acronyms of PAIR and MAGIC have recently been used to describe attributes of useful Mobile Internet applications. We can test possible Mobile Internet services against the principal that if a service is Personal, Available, Immediate, and Real time it has a good chance of being successful. Obviously these attributes don't only apply to Mobile Internet services. Between us we found many industries that can use this method, the personal services industry is a prime example as also is the fixed Internet. So PAIR is not specific to mobility and is unlikely to identify the early growth services in UMTS networks.

MAGIC

NTT DoCoMo believe that services have to be MAGIC to be successful. Mobile, Anytime, Globally, Integrated, Customised. The MAGIC formula provides a good indication of migrating fixed Internet services to the Mobile Internet, or for creating new services onto the Mobile Internet. Three of the MAGIC attributes we consider are stronger that the other two. A successful Mobile Internet service has to be Mobile otherwise it is not a Mobile Internet service. Users

3 Vignettes from a 3G Future

Betting on the Overtime Result

I am watching an exciting football game at the stadium. The game goes into overtime. Then I receive a message on my UMTS phone which gives me the chance to bet on the outcome of the overtime. Of course being a loyal fan I bet on my team scoring. The best aspect is that I can bet without having to leave my seat to go to a betting booth, and not risk missing any of the exciting overtime action.

The betting on the overtime and similar live game large audience betting situations are almost impossble to arrange with any other means except the mobile phone or other similar device. The betting facilities at a live game would be totally overwhelmed if over 10 000 people suddenly rush to place additional bets if a game goes into overtime. Also many people would not believe they could make it back to their seats in time to see the result, leaving the betting opportunity unused. If advertising, event information and betting opportunities are to be used at stadiums with thousands of people in attendance, a prudent operator would deploy extra UMTS network capacity to handle the projected traffic peaks at the location.

will want access Anytime and people want Customised or personalised services. But many services will be local and not global. Services that are big in Asia may not see the same take up in Europe or the Americas and vice-a-versa. The cultural aspects of UMTS service creation have to be carefully considered since the trusted McDonalds formula of one experience for all users will not work in the future.

UMTS services also don't have to be integrated, some will need to be like banking where integration of bank account details, PIN numbers and credit card details are all relevant to your financial transactions. Yet a joke service does not need to be integrated. Personally we would like to receive Dilbert every morning at 8:15. This is not integrated with anything apart from our personal view of humour in the office.

4P's

Another approach is to use the 4 P's: Portable, Personal, Processor, Proactive. (TIM) Telecom Italia Mobile, the largest GSM operator in Italy have proposed this as an alternative service definition criteria. Like the above methods this can also be used in many cases to offer some indication of UMTS services that could appeal the UMTS mass market.

Zero-One-Two-Three (0-1-2-3)

Recently Ericsson has come out with new mobile phone handsets which have been created with '0-1-2-3' by which the service has no written manuals on paper, has access to the internet behind one button, no service to have a service delay or lag of more than 2 s, and all menu choices are not longer than three keystrokes.

2.5 At last on attributes

With abilities to take advantage of location information, do multi-session work, provide timely services, and have the ability to generate income from micro-payments, new and exciting content will emerge. The attributes described here are useful, but a holistic approach specific to UMTS is developed in the next chapter, where the 5 M's (Movement Moment Me Money Machines) of UMTS Services are described. Still the attributes in this chapter can serve as launching points into new services, so we leave the combinations and applications to your imagination and offer you Carl Sagan's guidance: "Somewhere, something incredible is waiting to be known."

3

'I don't know the key to success, but the key to failure is to try to please everyone.'

Bill Cosby

The 5 M's of Services for UMTS:

Killer wanted

Tomi T Ahonen and *Joe Barrett*

If there is one question that we have been asked time and time again it's '*What is the killer application in UMTS?*' It is almost as if people have to have one thing that they can hang their hat on and say 'this is how the UMTS investment will be recovered'. Maybe it's because there have always been killer applications in the past, just like word processing and spreadsheets were the killer applications for personal computers, likewise e-mail and web browsing drove the growth of the fixed internet. In GSM (Global System for Mobile Communications) voice has been the main application with SMS (Short Message Service) or text messaging growing as an important revenue service. Certainly in countries of high mobile penetration the youth market seems to use the phone mostly for messaging, and thus SMS messaging could be seen as their killer application. Some could argue that pre-paid subscriptions could be considered as the killer application in GSM overall, since pre-pay has stimulated the growth into the mobile mass market Figure 3.1.

The 5 M's

MOVEMENT - escaping place (local, global, home-base, mobile)

MOMENT - expanding time (multitask, plan, postpone, stretch, fill, catch up, real-time)

ME - extending me (personal, relevant, customised, community, permission, language, multi-session)

MONEY - expending financial resources (m-commerce, micro-payments, m-banking, trusted partner)

MACHINES - empowering devices (telematics, machines, appliances, robots, metering devices, remote monitoring cameras, connecting with any conceivable machine)

Figure 3.1 The 5 M's.

With the Mobile Internet it is unlikely that one or a few global killer applications will appear soon and if they do, by the time we spot them they will, more than likely, already be well established. Furthermore we believe that one definitive killer application for the whole population is quite unlikely in the UMTS environment. An analogy could be drawn from television programming. On TV we do not have one killer application, but rather the selection and variety is the 'killer' and we buy television sets, cable and satellite service subscriptions, VCRs (Video Cassette Recorders), DVD (Digital Video Disc) players and other gadgets based upon our personal preferences. Someone wants sports, another soap operas, another comedy, yet another is addicted to news, and someone else wants movies. Most people are not even limited to one type of programme, so I might watch sports, news, light comedies and the occasional movie. What is then my killer application?

This does not offer much help for UMTS service creation managers. Based on the past experiences and an ever deeper understanding of how and why people use their mobile phones we can develop tools to create better services. The previous chapter has already explored several service creation formulae which can be of assistance in making mobile services. We will now offer a new tool which is designed specifically to assist in identifying and creating better mobile services.

3.1 The 5 M's of UMTS service definition

We have defined a method that we believe could provide a good insight into possible revenue generating UMTS applications or at least those applications or services that have the greatest chance of success. We call it the 5 M's (Movement, Moment, Me, Money and Machines). A successful UMTS service should meet all these attributes in various ways if it is to have any chance of becoming an early adopted service for the mass market. And in UMTS we are talking about the mass market being the main focus for operators in order for them to create fast market growth and early time to profitability. Some of the 5 M's may be more relevant to certain services than others but all should be apparent in some way if the service is to hit the growth curve we are looking for. The most successful services should be those that are strong in all 5 M's but it is important for all operators and UMTS service creation managers to remember that cultural differences in user behaviour will effect how each individual service is rated in each country.

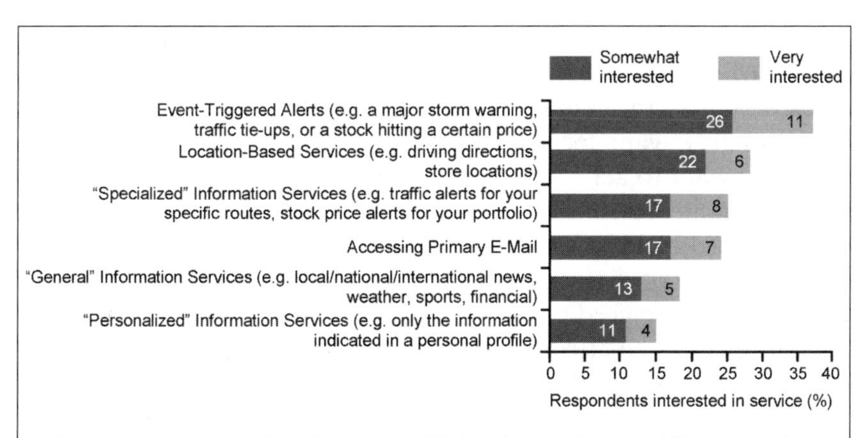

There are a multitude of studies into the likely take up of new mobile data services some carried out by leading analyst companies and others commissioned by operators and even vendors. It does appear from much of the research that there is a strong possibility that take up will be stronger than many pundits think. Yankee Group have many ongoing studies and this example shows customers interest in a number of service types.Source: Yankee Group Wireless/Mobile Services Report, Volume 2, Number 9, July 2001.

Movement – escaping the fixed place

A Mobile Internet service by nature must be mobile and it must enable the free movement of not only the user but also the service. Today, roaming is a major advantage of GSM and a profitable service for all these operators. People have come to expect that the services they subscribe to are always available when travelling in foreign countries, even if it is only when they take their summer vacation. The VHE (Virtual Home Environment) initiative started by ETSI and now part of the 3GPP (3G Partnership Program) should ensure that access to individual UMTS services is consistent no matter which country the user is in. This is what users will expect, nothing less. One aspect of the Movement attribute will be location. Location will often be an integral part of the service to the point that without it the service would not exist. Positional information is one of the key knowledge factors that Mobile Internet operators will use to add additional value to their mobile service offerings.

Movement services need to be intelligent in their handling of this location information. If you are visiting another country, and looking for a cash machine, it should offer you the locations in that city, not your home city. But if you want to get a stock market update on your portfolio, you want your own country stock market, not the local stock market of the country you are visiting. In the case of TV listings you probably want local listings, but if you are flying back home today, or want to activate your VCR remotely, you want easy access to your home town TV listings as well. So Movement attribute services need to behave intelligently for the user and in many cases offer a number of choices that will assist the user in quickly finding the information they need.

Movement as an attribute is the most defining aspect from mobile/cellular networks. With concepts such as roaming and remote access, this is the part of the 5 M's that the current mobile network operators know the best. This attribute is what sets the mobile services apart from the fixed services that most people experience today. Any service which is not high on the Movement dimension, is easily copied on other technologies, and conversely, a service which is high on the Movement attribute, will be difficult to replicate on competing technologies.

Moment – expanding the concept of time

We are all busy, even our kids don't appear to have time to talk to us anymore. We live in an instant world with fast food, instant cameras, 1 h film development, 30 min pizza delivery, 15 min hair cuts and 5 min shaves. So a UMTS service has to be relevant to the time of day or delivered at that crucial moment. The service may even check your calendar before it delivers the content or service to make sure that you are in a situation that is suitable for receiving the content you require. Undeniably users will define when and where they will use services or be pushed information. A Mobile Internet 'personal advert' is a good Moment service. I will want to know when someone is interested in purchasing my old sofa and what their offer is and I want to know the instant someone is selling a VW beetle in good condition for under 2000 dollars.

Services which use the Moment attribute allow for postponing, rescheduling, and last-minute behaviour, as well as to catch up on lost time. Busy people often use their mobile devices to make last-minute modifications to plans, or to leave a decision to the last moment. Technology such as voice mail, e-mail, SMS, etc., enables us to receive messages when it is convenient, thus effectively moving time or actually by moving communication to another point in the future. Manipulating time also sometimes works backwards, for example having easy access to yesterday's news.

Moment is the most activating attribute of the 5 M's. If the need is exceptional, say it is raining and you need to know where to buy an umbrella, the Moment attribute can result in an almost compulsive action by the user. It is very important to understand the Moment attribute, and to create as much opportunity as possible for the user to fulfil momentary needs and wants. The far-sighted operator will train its users to use the Momentary features to create almost addictive uses of the Moment services. The service portfolio should also collect impulses from users, who have momentary needs, to identify these occasions. The service portfolio should also prompt the user to seek for assistance from the mobile phone – in any urgent need, the answer is in your pocket.

Me – extending myself and my community

We don't want your services we want our services. We want what we want not what you want. The Internet and society is teaching us to become demanding consumers and more efficient executives so it will be my content, my services, what I want, what my interests are, what supports my work needs, what makes my life easier. Yet though we are all individuals we also have communities that we associate with. We need to be part of society. We need to feel loved, respected, listened to and to have a sense of belonging. It could be our family, friends, business associates, tennis partners, a fan club, horse riding society, pressure groups, political organisations, the list is endless.

The Me attribute is by nature also the most complicated to understand because it includes our tastes, likes, dislikes, image, style, views, values, intelligence, impressions and feelings. Get it right and you can create a powerful winning service. Get it wrong and the service will die of neglect. Ring tones are a perfect example of a service that hits at the heart of the Me attribute. It is one of the things that make our phones personal to us and expresses our image, likes, tastes. To successfully address this attribute with a new UMTS service, operators will have to create a one to one relationship with their users. This will only happen with extensive user profiling where behavioural and usage patterns can be built up and disseminated so that individual customer wants and needs are better understood.

It is also important to understand that the Me attribute is an extension of myself into my communities. So services that help us keep in contact, that help us build and share are all addressing the community aspects of Me. At work a group of colleagues who work on the same project could be such a community. They can communicate effectively via a number of devices, fixed and mobile to achieve better productivity. If the members develop their community further, they might go out for some beers after the launch of the product, or after reaching some important project milestone. The UMTS mobile terminal offers numerous ways to build and maintain that community by allowing communication, contact, caring and sharing between members. Similarly a fan club of a rock band might share sound clips, pictures, and receive fan mail via their UMTS terminal. A family is the most natural of communities and mobile phones today are playing an ever expanding role in family communication as we are all more mobile and have less personal family contact on a daily basis than we did 20 or even 10 years ago.

Groups or communities are now becoming very popular on the Internet. Yahoo is probably the most well known. See www.groups.yahoo.com. You can join specific groups according to your interests or make up your own. They are ideal for virtual communities like study groups or virtual teams when the members are dispersed around the country or even around the world. The next step is for these groups to go mobile so that you can post pictures, files, news etc. and everyone in the group has access to it. This kind of Me service enables any number of people with the same interests or needs to communicate.

The Me attribute is the attribute which the individual user will value highest. It is a reflection and extension of the self, the ego, the person, as well as the person's connection to the community and a reflection of their image in it. The operator should build a strong 'I understand you' feeling into its services. The Me attribute is also that which is easiest to hurt by a trusted friend. The more the operator develops a close relationship with its customers, the more those customers are prone to become upset if their information is used against them, or without their express prior approval. Developing the trust between customer and supplier is needed. Honestly listen to your customers, and deploy services which can respond to individual personal needs.

Money – expending financial resources

A UMTS service must generate money. This is so obvious we are going to say it again. A UMTS service must generate money. Yet it is not as simple as taking a current Internet service or high street shopping experience and delivering it via the mobile device, expecting it to be the most wonderful service a user has ever seen. A UMTS service must deliver value. Merrill Lynch in their Wireless Internet report stated that: "We believe the value in the Mobile Internet will not lie so much in the content but in how the content is cut and distributed[1]" It is also probable that over 90% of the current Internet content is not suitable in its current form for the Mobile Internet since it is based on pulling the user to the site and pushing the information to the user via an activated application.

Operators will have to create what we like to call Invoiceability into their applications and services. So if you can't bill for a service – kill it. Invoiceability also may not reside in the customer part of the value chain for all services. The cost of delivering a mobile promotional message to John that Britney Spears is booked to play at the Hartwall Areena in Helsinki in three months time may well come from the stadiums advertising budget for this event. Of course receiving the promotional message would not cost John anything. When John purchases his tickets to see Britney via his UMTS terminal the UMTS traffic cost could be wavered since the operator will make revenue from the ticket sale. So John would not pay anything extra for making his purchase via his UMTS terminal. The price could be the same as standing in line at the ticket counter to buy the ticket. The utility in that case is convenience.

Services around Money will also include actual payments by the phone user to the operator or payments for content, such as paying for playing a game or downloading music. In fact we believe that the UMTS mobile device has the greatest opportunity to make society almost totally cashless. SMART Communications in the Philippines already have such a service and this is being delivered using SMS as the bearer technology.[2] So as we have already discussed many services will start before UMTS.

Credit card companies are not eager to embrace small sum payments. Yet the mobile operators are ideally positioned to accept micro-payments for almost anything since they already have the billing systems and processes in place to handle these kind of transactions. By exploiting this advantage UMTS operators can develop their total service offering to the point where the mobile phone bill becomes more than just payment for telecommunications services. The Money attribute has to be considered for every UMTS service.

[1] Wireless Internet, Merrill Lynch June, 2000.
[2] http://www.smart.com.ph/Main.asp.

4 Vignettes from a 3G Future

Lend me a few dollars

I get a call from my nephew visiting our relatives in a small town in the countryside. He has lost his wallet and is without his money and now cannot get back home. He does have his UMTS phone. As he is young, his bank account is pretty close to zero, so he cannot use his mobile banking. But he remembers that I have told him in an emergency to call me. I take my UMTS phone, and select from my electronic banking services the Send Money Now option and send him immediately 50 dollars. He uses that money to pay for the bus ticket and a snack along the way, and gets safely home.

The UMTS terminal will become a trusted partner for all who carry it. And exactly as has happened with the GSM phone, the owner of the UMTS phone will insist on carrying it everywhere. One of the most convenient features of the UMTS phone is its trustworthiness. It will be as personal as credit cards and banking cards and soon will probably be as widely accepted as a payment device as credit cards are today. One of the new ways that the UMTS phone will be used is to lend a friend a little bit of money, exactly like lending a 20 dollar banknote. Similar to "wiring" money, or sending a check, or paying by bank transfer, but much more convenient; sending money from your account to the friend will be as easy as sending an SMS is today.

The Money attribute is the one valued most by the service partners. It is the magic key to profits. The UMTS operator should enable all types of Money features and help make any transactions possible, with options and features to provide transaction histories and billing information etc., as parts of value-adds to consumers, and as billable services to partners and content providers. The micro payments option is the most dramatic single element of the UMTS environment, and it will enable the migration of content from the fixed Internet into the UMTS environment. Micro payments are the components which eliminate the chicken-and-egg problem with most emerging technologies. The content tends to exist on the fixed Internet, only it is not making any money there. The micro payment possibility with UMTS allows that content to generate revenue. The best content will follow the money trail, and the users will follow the best content.

Machines – empowering gadgets

UMTS mobile gadgets come in all shapes and sizes. Just like the car industry has moved on from Henry Ford's famous any colour as long as its black quote, the mobile phone industry has moved on from black or grey covers to where the mobile phone is becoming a lifestyle accessory, changing with our moods and whims. We now have a variety of mobile device segments. Low end, fashion, premier, youth, business, etc. In UMTS these will extend to home, office, car and many more areas as we use a variety of devices to access our personalised content depending on where we are and what is the ideal device. For a UMTS service to meet the Machine attribute there must be mobile devices and sometimes non mobile devices that can provide satisfactory delivery of the service or content. A good example of this is mobile video. If a proposed mobile video service is built around high quality screen resolution on the mobile device and current screen technology can not provide that resolution level then the service will fail because it will not meet the users' expectations. So when we talk about machines we also mean the mobile device as well.

With any mobile service the terminal will always be the gating factor. The first GPRS (General Packet Radio System) terminals were based on text and simple graphics form factors so the complexity of the service and the user experience is then limited by these devices. Creating a service that is based on imaging or needs colour for effect is not going to succeed with these terminals. Even if there are terminals that have the attributes to deliver the service, the cost of the terminal may restrict the take up, for instance if a service directed at a youth segment needs a premium terminal it is unlikely to succeed since the target audience may not be able to purchase the UMTS terminal needed.

The Machine attribute can also reflect future interaction with remote devices and even device to device communication such as the telematics of an automobile. Your car will soon communicate with the car manufacturer or service company to warn about possible defects or service needs. But Machine attributes go much beyond this. Some appliances and devices are already in existence which can be controlled remotely. For example VCR's, washing machines and refrigerators already exist which allow this kind of control via the Internet. The natural evolution is that such devices can be accessed from terminals on the Mobile Internet. So the UMTS device itself can be a normal UMTS phone, accessing another device which is a machine. Such services are too numerous to list. Open the yellow pages and start counting.

The Machines attribute will connect and enable a population of UMTS mobile network devices which will dwarf the human population. Early estimates suggest connected devices will soon have twice the population of humans, and later in the UMTS network life cycle the amount of non-human users might be ten times that of people. Some of the machines will produce only small amounts of traffic, but others might produce more traffic than the most productive humans. The Machines aspect is the biggest potential area for growth in UMTS services.

The Machines attribute is a key for better profits in UMTS services. The ability to automate data delivery processes, to address and access global user pools, and build scale to even the most fragmented markets, allows for better profitability than any currently existing alternate delivery mechanism of billable digital content. The operator should not underestimate the power of the Machines attribute.

3.2 Testing the 5 M's: the Mobile Ring Tone

Lets validate our 5 M's against one current yet simple service to see if this method can offer some guidance to successful UMTS services in the future. Let us examine the Mobile Ring Tone, which some call Ringing Tone, in other words the various short songs that can be downloaded for the mobile phone to play our favourite tune when it rings.

Each one of the 5 M's can be given a score from 0 to 5 depending on how well the service fits to each 'M'.

Movement: Rating 5. I can download the Ring Tone from wherever I am and it is simple.

Moment: Rating 5. I can select the moment that is convenient to me to do the download. If the Ring Tone is promoted on the radio, TV or newspaper I can chose that very moment to download it.

Me: Rating 5. The Ring Tone is based on my tastes and likes and it says something about the kind of person I am. If it is unique amongst my friends then I will like it even more. I can forward it to my friends and thus connect with my community.

Money: Rating 5. I don't mind paying for the Ring Tone because it is quite inexpensive, in fact I change my Ring Tone every 1 or 2 months since I get bored with it or maybe I hear it on lots of other mobiles.

Machines. Rating 3. When I download, I do so interacting only with a machine (server). My mobile plays the Ring Tone quite well but I would like CD sound quality.

Total Service rating 23/25.

Any service which brings high value through any one of the 5 M's has potential for Mobile Internet success. The more of the 5 M's that are highly relevant, gaining a high score, the more likely the service becomes particularly suited and beneficial to mobile customers. The high rating services can be considered to have initial growth potential.

The 5 M's

Movement – escaping place (local, global, home-base, mobile)
Moment – expanding time (multitask, plan, postpone, stretch, fill, catch up, real-time)
Me – extending me (personal, relevant, customised, community, permission, language, multi-session)
Money – expending financial resources (m-commerce, micro-payments, m-banking, trusted partner)
Machines – empowering devices (telematics, machines, appliances, robots, metering devices, remote monitoring cameras, connecting with any conceivable machine.)

3.3 Using the 5 M's

The 5 M's are one way to identify the attributes for successful services for UMTS especially for initial services. The 5 M's can help guide service creation managers and teams in designing services that are desirable, valuable and 'sticky'. The 5 M's should not be considered in isolation, and no service will be optimised if it is using only one of the five dimensions.

The merits on any of the five attributes are also relative to competitive offerings in the marketplace, both on other UMTS services and via other means. This means that the operator and service provider should constantly keep an eye on what the competition is doing, and map the competitor offering on the range of the 5 M's. The marketplace for mobile services will not know what is a theoretical maximum performance on any given attribute, but will be comparing services against similar services. The relative superiority on any of the 5 M's will be what determines whose service is the best.

It is good to remember that each of the 5 M's speaks to a different need in the overall marketing and consumption of mobile services. Movement is the defining attribute which differentiates UMTS services from those delivered on other technologies. Moment has the highest activating power of the attributes. Me is the most personal and one which is seen most valuable to the user (or community). Money is how UMTS services attract best content. Machines is how UMTS services can be made more profitable.

3.4 Finally on the 5 M's

We have found the 5 M's to help us in our understanding of which services might prove popular and how to make any given service *more attractive* in the UMTS environment. The 5 M's are but one tool in this search for the evasive killer application or killer cocktail and of course many other tools exist or will be developed as we all search for the secrets that will lead to a profitable UMTS business and make the future happen.

For success in UMTS the UMTS operator and its partners must deploy attractive services which will bring value and utility to the user. The UMTS operators will be providing those services in a heated competitive environment where many of the world's biggest corporations are fighting for their piece of this new cake. There will be winners and there will be losers. An ideal service would be one which is very fast to spread within single and multiple networks and one which the competition is not able to quickly copy. The ideal service should be priced low enough to get massive adoption, but still high enough to bring solid profits. This requires a great deal of creativity and imagination and the ability to think out of the box and beyond our own experiences and perceptions. In the end it comes down to winning and change, on which Anthony Jay wrote: "Changing things is central to leadership, and changing them before others is creativeness."

4

'Laughter is the shortest distance between people.'

Victor Borge

Services to Address Movement Needs:

Escaping the Fixed Place

Päivi Keskinen, Michael D Smith and *Tomi T Ahonen*

In the following five chapters we will look in more detail at what we mean by the 5 M's (Movement Moment Me Money Machines). We start in this chapter with the first of the five attributes, Movement. The ability for a mobile service to escape a fixed place means that the service will seem perfect regardless of location and whenever relevant, specific to location or mobility. The main types of Movement aspects are locality, globally, home-base, and mobility. Locality means that the service provides specific location information. If you are visiting another city, and ask for movie listings, the service should provide local movie theatre listings, not your home town listings. Globally means that some services need to provide the same service world-wide. The home based service means that it knows where your home base interest is, and provides automatically, or upon easy request, the home town information. If you know you are returning home, and want to know what is on TV, you'd want your home town TV listings. Mobility means that the service transfers with you as you move. And services need to move across the network at the highway speeds of moving automobiles. The most obvious service is for automobile drivers, looking for the nearest hotel or place to eat.

When designing a new service, the operator or service provider should try to include as many of the 5 M's as possible. In addition to Movement, the other 4 M's are Moment, Me, Money and Machines. In some cases the service might include several elements from one M, and very likely variations will be made of similar services by varying the degree within one or more of the 5 M's.

The services described in this chapter have a high benefit on the Movement attribute. There are hundreds of such services and we have a chance only to explore a few of them. To illustrate by way of example some of the obvious uses of Movement type services, they include:

Access to internet
Access to WAP (Wireless Application Protocol) content
Advertising location push (vignette)
Business access to intranet
Business data transfer
Business project management
Calendar synchronising personal
Collaborative applications at the workplace
Consultations with medical doctor
Dictionary
e-mail
e-mail to voice (vignette)
Emergency medical care
Fax
Info tourist guide
Info traffic
Internet browsing
Local bus, subway etc schedules info
Local restaurants
Local weather
Location detection 'where am I'
Location guidance 'where is it'
Mobile government apply for permits
Monitoring patients
Order taking
Organiser synchronisation
Rescheduling flights
Secure connectivity at workplace
SMS (Short Message Service) text messaging
Telehealth (telemedicine)
Tourist assistance
Traffic info
Travel guidance
Travel holiday packages
Travel hotels
Voice home zone

This chapter will discuss some services which have a strong benefit from the Movement attribute. The services are not offered in any order of importance, nor do we attempt to cover the subject matter comprehensively. A deeper discussion of a few services is useful for a clearer understanding of the Movement attribute and how services for UMTS (Universal Mobile Telecommunications System) can benefit from it.

4.1 Adding value to travelling life

Perhaps the most obvious area of services where the Movement attribute is significant, is that of services around travel and tourism. Many of the current services and the underlying travel needs can be recognised from current services for travellers. These range from various books and guides printed for tourists, to internet services around tourism, to business travel services, onto frequent flier programmes and so forth.

Travellers both tourists and travelling businessmen, have needs quite different from the people who live in the travel destinations. Travellers typically need to find hotels, restaurants and cash machines. They want to know exchange rates and time differences. Travellers have personal interests such as wanting to visit museums, catch the cultural events, visit night spots and local entertainment. Travellers often have sudden changes of plans and services to help last minute bookings, changes, updates., etc. can be very useful. A separate category is services related to travelling by automobile, which will be discussed in depth in the Machines chapter later in this book.

Where is it?

Travellers have particular needs when they arrive at a new place. Any of the locating needs – where is the hotel, cash machine, restaurant, car rental, etc are obvious candidates for UMTS services. These services benefit from location information and personalisation information, and many simple services to assist travellers are appearing around the world today. Maps, guides, local entertainment listings, etc are also naturals to benefit from the Movement attribute. With the 5 M’s any services can be enhanced to improve on what is available currently in printed and electronic forms, such as offering personal preferences in restaurants, and adding click-to-order abilities to train ticket information requests.

The service developer needs to note that today's travellers are becoming ever more demanding in the kind of services they want. Many industries are serving these needs, and the fixed Internet is providing many different kind of services to this segment. Applications and travel services for UMTS need to be highly focused on a person's personal lifestyle to be perceived as valuable. Nobody is going to want to simply surf the whole fixed internet using a UMTS terminal in search of the ultimate information relating to what may be a short one time visit to one country. Every data application used should add value and make the journey more enjoyable and more fun.

Rescheduling flights

When services relate to Movement, the UMTS service should anticipate the next need. For example, a service that allows a traveller to look at an airline's schedules on the way to the airport and then be told that they are going to miss the plane because they will be late is OK as a service, but in reality is rather negative for the user who probably already knows they are going to miss the plane. What really adds value is to be offered the best alternative flight, its cost, departure and arrival times and to be able to change the booking there and then, using the mobile terminal.

Intelligent time tables

Services need to recognise the immediate needs relating to that location. *Where* is the nearest bus stop? *Which* bus goes to my destination? *When* will my bus arrive? This kind of information already exists in many places for access via the fixed Internet or for example via display units at the bus stop.

Operators can provide entertainment for the time that it will take for the bus to arrive, and even an enterprising taxi service might use a push promotional service to all who ask for a bus arrival and the service says it takes more than, for example 15 min. It may be that such a commuter would normally not consider a taxi, but if one was offered they would actually take the taxi option if they could be guaranteed it would arrive before the bus.

Services relating to travel are numerous and most of them can either be provided via UMTS or can be enhanced with UMTS. The operator needs to also keep in mind that most travellers have a heightened need for the services they are seeking, thus they are less sensitive to price. The operator should make sure that its service roaming is set to accommodate the traveller who, when compared to the average mobile phone user, already tends to spend considerably more on the current mobile phone technologies. When specific tourist and traveller oriented services become available on UMTS, this trend is likely to become even stronger where tourists may become the most lucrative user segment.

4.2 Business to employee (B2E) services

Most services sold to residential customers also exist in a very similar form on offer to business customers. There are, however, some services that are used primarily only by businesses and not by private citizens as consumers, an example of some of these and their penetration rates can be seen in figure 4.1. Business applications may be split into three broad categories of B2B (Business to Business), B2C (Business to Consumer) and B2E (Business to Employee). This section focuses on B2E and B2B applications, where the Movement attribute offers the greatest benefits in a business communication context. Business to Consumer applications are dealt in more detail in the Money chapter later in this book.

The business to employee segment is itself a complex market, with many applications delivered over a number of possible architectures and via several business models. As organisations are adopting e-business means of collaborating, sharing, team work, outsourcing, and empowering their employees, most internal processes are in turmoil and being redesigned. As access to information is redefined, usually broadened, then this brings about new challenges for the organisation on how best to handle the information. The enterprise needs to consider how to handle the information needs and decision making with part-time workers, outsourced workers, partners,

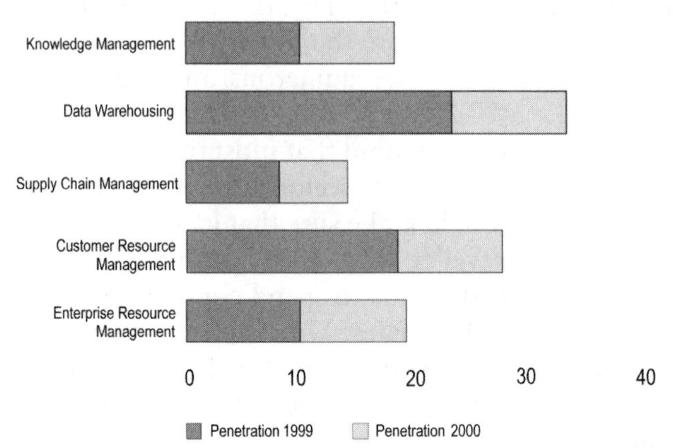

Figure 4.1 An example of common B2B applications and their penetration rates. Source: IDC, Global IT Survey 2000.
There are a number of growth areas in business applications that can be targeted by the UMTS operator. The ones that will be of greatest value to all parties will be those that create better efficiency in business processes.

subcontractors, etc. Who has access to what information, who can control processes and make decisions, etc. These all have profound impacts to how the employee of the future will perceive work and participate in it. The Movement attribute brings particularly strong impacts to the work situations as employees can work part-time, from home and on the road, and at differing office locations.

It is important to consider the level of integration between the application and other business applications and processes. As technology evolves, often an innovation is introduced first as a stand-alone system to improve one given process often in one department or business unit at a corporation. Then, as the technology is adopted more widely or its benefits can be seen to other systems and processes, the SI (System Integration) element of technology implementation will need to be addressed. This was for example with e-mail. Very early e-mail was only among mainframe computer terminals. Its spread came to PC based LANs (Local Area Networks) which allowed dozens of office workers within one office or floor to communicate with each other. Then LANs and e-mail systems were connected to each other, and eventually to the Internet, which brought about a much greater utility to the technological innovation.

Today the integration is taken several steps further by integrating the e-mail to mobile access and SMS text messaging. When the UMTS operator or application developer introduces a new innovation, it may start out with a stand-alone solution for one department or unit at a corporate customer. Once a relationship is established and the customer culture identified then additional applications may be deployed which require integration work, and which may impact on business processes. It must be noted that some applications may be implemented as stand-alone, and can then be integrated with other applications. An example of such an application is PIM (Personal Information Management), which at the simplest level is an electronic calendar for the individual. PIM can be integrated with other systems to provide the intelligence of where staff are forecast to be at specified times. This information may be used in a task management system.

Secure connectivity

In order to provide mobile access to applications a secure connectivity solution is required. This will ensure security from the terminal, through the remote access network to the applications servers. Any connectivity solution must protect corporate data from access by unauthorised persons. The requirements of any application may be categorised as follows:

- Confidentiality and Privacy – no one can intercept information
- Integrity – information is not altered during transmission
- Authentication and Identification – the person/system is actually the one they claim to be
- Non repudiation – messages once sent cannot be denied by the sender
- Access control – access to a system is restricted to parties who have been granted permission

Not only personal or corporate information must be secure from loss, corruption or theft, but so too must any related management information, usage information and charging information. Information must be secure from attack by other remote users, from external networks, operator personnel, and even from the effects of malfunctioning equipment.

Internet and Intranet access

UMTS networks are particularly suited to providing Internet and Intranet access to mobile users. Most new networking applications will be built on Internet technology, specifically the IP (Internet Protocol) protocol suite used by services and traffic on the internet. The basic IP protocol provides no real security protection and many applications require additional security layers, built on top of the underlying IP protocol. Security is achieved via a combination of techniques, including physical security against network element access, host authentication, encrypted links, personal security with authorisation, and digital signatures/certification authorities.

Security technologies alone cannot guarantee security. Total security is determined by the weakest link of the chain. When considering a security solution the operator must look to employee awareness and training, and the continuous auditing, monitoring and updating of security policy and processes. A company should define a complete security policy involving all relevant departments, such as Human Resources, Board Members, Information Management, Sales, Product Development.

Collaborative applications

Applications which are already widely used by enterprise employees such as messaging and calendar functions may be significantly enhanced by offering improved UMTS access. Messaging systems in current use are fragmented, with users accessing SMS text messaging, e-mail, and possibly even separate fixed and mobile voice mail systems. Unified Messaging services offer to bring these different media together as one manageable service, to convert between different message formats and to offer messages to the enterprise user in the format most appropriate to the terminal used. Unified messaging is discussed separately later in this chapter.

Many handheld business terminals will offer a set of personal applications such as: Diary Manager (Calendar), Task Manager & Action Chaser, Contacts & Address Book, etc. In the UMTS world these functions will support Voice, Image and Video Entries. A Network virtual PDA (Personal Data Assistant) will add 'Multi-

Person' Personal Applications, for example 'Meeting Arranger', 'Travel Organiser', which use personal calendar information from multiple employees. Resilience of terminal data may be provided via central automatic backup systems and solutions. Users will require accessibility to common information from both Fixed and Mobile terminals.

4.3 Business to business (B2B) services

More complex applications linked to business processes include CRM (Customer Relationship Management), ERP (Enterprise Resource Planning) and SCM (Supply Chain Management). These applications have more of a B2B focus or are refinements of the very business core processes. They tend to require business consultancy and system integrators in large and expensive projects which deliver tailored solutions spanning business processes and technology. The ASP (Application Service Provider) model allows a generic application set to be developed and offered to businesses which would not consider a tailored solution. As an example of this type of application, CRM is discussed in further detail below.

CRM (Customer Relationship Management)

New communication and application technologies enable businesses to move away from the traditional structures of Human Resources, Sales and Marketing, Accounts, etc. CRM systems allow the adoption of a totally customer-centric method of working. CRM is not an easily implemented solution. It requires a fundamental change of processes throughout the organisation.

CRM systems enable a business to capture customer information from all channels of contact, building a comprehensive picture of customer behaviour and requirements. A CRM system with mobile access capability will enable a business to see when and from where a customer accessed a web site, as well as the products and services browsed, purchased, and any particular requirements specified. This information can be used for targeted marketing, resulting in both increased customer satisfaction and higher sales. At a time when

5 Vignettes from a 3G Future

We Have Your Favourite Scotch Whiskies on Sale

I visit Edinburgh for the first time, and walk on the street, and receive an advertisement on my UMTS phone. It says that a whisky store has single malt scotch whiskies on sale, and I am very close to the store. I receive an electronic discount coupon I can use on my first purchase. The store gives me a map of the immediate few blocks and I can see I am within 3 minutes walking distance of the store. I notice I have some extra time and go in. I like it that as I don't care much about vodkas, cognacs, beers or wines, I never get those advertisements, only ads on what I really like, and in my case it is scotch whisky.

The "permission push" advertisements work best when the actual advertisement received addresses a real and personal interest. Combined with location information, the targetted permission push ad becomes an extremely powerful and activating marketing tool, and one which is currently almost totally unavailable in any other media . If the target is a tennis fan and thus interested in sports, sending sports ads would not be precise enough. The person is not necessarily interested in football, golf or biking, only tennis. But for a true fan of tennis, receiving tennis-related news, information and advertising would generally be very welcome. And for the advertiser of tennis games, equipment, accessories, etc., sending such ads to an avid golfer, who does not care for tennis, would be a waste of marketing effort.

companies are finding it harder to differentiate themselves with their products or services alone, CRM enables the relationship development with the customer to become the differentiating factor.

A CRM portal may be used to enhance customer interaction at one of four levels:

1 Provide information
2 Alternative sales channel
3 Customer's purchase aid, e.g. product specification tool
4 Interactive management of customer relationship

CRM has the potential to increase sales revenue, whilst at the same time reducing the length of the sales cycle, reducing the cost of sales and increasing customer satisfaction. The opening up of the Enterprise CRM application to customers or the enterprise sales staff by offering mobile access can bring dramatic improvements in the business benefits achieved. Sales staff will have access to critical information whilst they are in front of the customer, allowing marketing material to be presented which is targeted precisely at that customer, and allowing ordering information to be captured and processed in real time as well as online.

The input of data into the CRM is also enhanced by mobility. If the particular information must first be 'stored' and then 'entered' into the system there is loss due to uncompleted tasks. For example when a sales representative visits a customer, jots down a note, and then enters the data back at the office via his PC, a time lag is experienced, and much more importantly, often the task gets forgotten because other urgent things at the office took precedence. When the UMTS terminal allows instant entry of the data to the CRM, then the sales representative can make the data entry immediately, for example logging a complaint right then and there at the customer site in the presence of the customer who made the complaint. Or if the sales representative does not want to address some issue in the presence of the customer, such as some details about the competition, then the sales rep can enter the data at the first available moment such as in the taxi on the way back to the office. The CRM is only as good as the data being collected by it, and if individuals are given too long to wait before data is entered, invariably a portion of that data entry work never gets done. It causes the expensive CRM solution to under perform.

4.4 Order entry

deally productive sales representatives would prefer access to perfect information systems from the office and yet be able to be at the customer site constantly. Typically today at any IT-savvy company, prior to the visit to the customer the sales representative will consult the customer database to obtain background material concerning the customer. The purchasing history may be referenced, and compared to forecast information. The online sales and marketing application may use this information together with current stock levels, and production information in order to formulate real time and personal promotional offers. The information available to the system and ultimately to the sales representative ensures that customer needs are met with maximum profitability.

A UMTS based system allows for even better integration of the corporate systems to the travelling sales representative. On the UMTS terminal a sales representative will have the ability to connect with the calendar, have schedules made and revised on the go, read e-mails and make contacts while travelling from one customer site to another. A big key will be the personalised portal set up by the employer's portal managers. Through the portal the busy sales representative can tailor the UMTS terminal interface to include only those information sources which are most needed and exclude all the ones that would create clutter and information overload.

Such a system would help in other ways. Comprehensive marketing and product presentation material would be available to the salesman whilst at the customers premises, allowing any questions to be handled on the spot. Ordering could be processed online, with no post processing back at the office. This would ensure that errors are minimised and instant confirmation is available to the customer. The sales representative could for example confirm delivery dates based on current production schedules and real time stock levels. Delivery times may even be shortened to reflect the earlier capture of the order by the production system.

This type of service would need a lot of careful thought on the data replication and IT sides of managing the information flow of the business customer. A prudent business customer oriented UMTS operator will develop this internal service creation talent, or get a strong partner and build business customer satisfaction by ensuring that its business portal services rank among the best in the industry.

4.5 Telehealth (telemedicine)

Telehealth, or as some prefer to call it telemedicine, is a means to provide healthcare and other medical services remotely through modern communications networks. Some basic objectives of telehealth services are to reduce treatment costs, to provide individuals with less expensive access to health related information, and to save people's time e.g. travelling to routine check-ups. The healthcare system is already in practice using telecommunications and Internet services for the transfer of patient and other related data between healthcare professionals. The next natural step is to extend the capabilities of communications networks to serve the contact between the healthcare provider and patient. Mobility adds value to telemedicine since consultation and treatment can be provided virtually in any location where access to a mobile communications system is available. This means significant improvements to emergency and home treatment, routine check-ups, medical consultation and sports medicine applications.

In spite of clear opportunities in the mobile telehealth field, the practical implementations so far have been few. One reason for this is that there are in fact a few fundamentally important issues in the transfer of patient data which are not completely ensured by the current (mobile or fixed line) technologies, namely the security and privacy of the data. Other factors holding back potential users have so far been the lack of meaningful and extensive content services and terminals. Within UMTS systems, user security features, user authentication, encryption and the user's traffic integrity aim to guarantee the performance of data transfer on the radio path and on the core network[1]. Compared to the current mobile networks, UMTS also clearly brings more possibilities for meaningful services with faster data connections for transferring the critical patient data, as well as its multimedia (combined sound, image and data) capabilities.

[1] UMTS Forum Report Number 9, 2000, p 43.

Mobile telehealth brings sizeable cost savings and flexibility to different stakeholders in the healthcare system. To hospitals, mobile telehealth means reduced bed stay and cost of care. To local health care providers, such as private practitioners, it means a greater reach of consultancy services. To patients themselves, mobile telemedicine means independence and fewer doctors' visits. Most of the routine check-ups, renewal of prescriptions, changes in daily drug taking programmes etc. can be managed remotely. Preventive treatment also becomes more effective when the clients have the option to send their vital parameters for the physician's consideration whenever convenient. In emergency care, mobile telehealth can save lives through immediate provision of the right instructions to paramedics.

From the need-base or functional point of view, mobile telehealth services can be divided to three categories: monitoring, emergency care and consultation:

Monitoring patients

Patients can be monitored at their home and on the move. Monitoring equipment can be connected to doctors or hospital computers on demand for further action[2]. Monitoring can be further divided into preventive care or home care. Preventive care may become especially popular in countries where the average age of population is steadily growing, for example most Western-European countries. Home care refers typically to the post-hospital monitoring period. Many simple mobile monitoring services are already in existence around the world. For the most part they are simple 'I am in trouble' panic button-type services, but these can be enhanced greatly with the advent of multimedia in UMTS.

Emergency care

In emergency cases, mobile communications is one of the most crucial instruments to ensure that patients are given correct treatment and medication during transportation to a hospital. First examples of remote assistance have been reported by the media, of a doctor advis-

[2] UMTS Forum Report Number 9 2000, p 43.

ing a nurse in an ambulance and saving a patient's life, etc. These isolated instances will grow into formalised procedures as the technology and its application become commonplace in the medical industry.

Consultations with medical doctor

The UMTS environment allows new ways for medical doctors to provide assistance to their patients. Consultation services allow patients to receive prescriptions, feedback information and advice from their doctors. Equally, and very importantly to make the service viable from the healthcare industry point-of-view, the UMTS environment will allow the patient to pay for the access to the doctor and the doctor's services. The ability of the UMTS environment to handle the payments involved will be a catalyst to bring healthcare services to the mobile world.

The business model of any given telehealth service is a complicated one, due to the large number of potential interest groups. In addition to the client or end user, there are three other main players in this scenario. The healthcare service provider is a business entity that is authorised to provide medical or healthcare services, such as a hospital. A new business entity can be called the 'mobile telemedicine (or telehealth) service provider'. It does not provide the primary treatment but takes care of all equipment, software and supporting services that are necessary to implement and run the mobile telehealth service concept. As a total solution provider, the mobile telehealth service provider has to co-operate with telecom network operators, biosignal[3] technology solution providers and system integrators. Payers in this scenario are expected to be insurance companies, national agencies or employers that have a role to pay for the treatment partly or in full, depending on the compensation model.

If they are not willing to enter the mobile telehealth service provision business, UMTS network operators may still generate revenues at least from the sales of new subscriptions (the healthcare terminals of the future may have their own SIM (Subscriber Identity Module)

[3] Biosignal technologies refer to the technical solutions used for measuring and transmitting vital parameters of clients, typically blood pressure, pulse rate, temperature, ECG (ElectroCardioGram) or glucose content of blood.

cards) and data capacity or airtime usage in mobile networks. The traffic generated by mobile telehealth services alone can be a significant new revenue source for the telecom operators. If we calculate conservatively that the penetration of mobile telehealth remote monitoring terminals will be 3% in 2005 in the UK and their average usage 30 min a month, price per Megabyte of data being 0.20 dollars and the average application throughput speed 43.2 Kbit/s, the yearly revenue generated by the traffic will be 41 million dollars.

All the technologies needed to make mobile telehealth a reality are already in place, and also practical services are starting to emerge. For example, in November 2000, Nokia together with the Finnish mobile operator Sonera and private hospital Mehiläinen in Helsinki Finland, launched a joint trial for mobile telehealth service in Finland. When UMTS networks are up and running, they will greatly enhance the possibilities of mobile telehealth by introducing higher application speeds, quality of service classes, multimedia applications and advanced security features.

4.6 Messaging

A most dramatic change in behaviour is caused by widespread use of SMS text messaging. Mobile phone users have found numerous new instances of communicating, where communications were either not possible or practical before; not even with the same mobile phone. These include such instances as noisy locations, for example a night club, instances when talking on the phone would be disturbing such as a meeting, and instances where you need 'secretive' communications such as children sending messages to each other during class. Sending SMS text messages is providing new opportunities for communicating where other means have previously been near impossible.

SMS text messaging

The 'texting' or SMS text messaging explosion has taken the industry by surprise, as on first impression the very service seems to be so counter-intuitive. After all, to use a current second generation mobile phone to type messages, one usually has to hit the number keypad numerous times for every alphabet character and for casual users it

6 Vignettes from a 3G Future

Greetings from the Whole Gang

My work buddies and I are out having a few beers after work. We are in our favourite bar and a couple of friends from the Singapore office walk into the bar, as they have just arrived for a meeting. We have a beer, and hear that one of our colleagues did not fly in from Singapore because his wife just had a baby. When we call him, we reach him at home and he turns on the UMTS video phone and shows us live images of his firstborn. We cheer with him.

Probably not all phone calls will be video calls. But in many cases after it has become very easy to establish the video telephone connection, the use of video in calls will become ever more frequent. A new phoning culture will probably evolve to handle video calls. Video calls will need to be real time video streaming, so their capacity use will be at high quality classes and operators will be able to price them accordingly.

seems to take forever to compose a simple short message. Each message typically costs about 15 cents and the limit in characters is 160. One would think that for most cases, calling a person and saying in a few seconds what would take perhaps minutes to type, would be much more practical and efficient.

That logic was based on making faulty assumptions based on previous behaviour, and did not take into account the new types of uses where subconscious needs had existed but could not be acted upon until now. To illustrate by simple example, let us assume I would find out that tomorrow's early-morning meeting was moved, and I only found out about it very late in the evening. I would want to tell my colleagues not to rush to the early morning meeting, but at that late hour I would not want to call them for fear that they might already be asleep. So before SMS there was no way to deliver the message. I would not call on the phone at a late hour, but today I can send the SMS text message. The message can unobtrusively deliver the important part of the message such as "Tomorrow's early morning meeting has been cancelled."

The growth in messaging has produced changes in behaviour in countries where mobile phone penetrations are over 60%. The changes in behaviour include last-minute changes in schedules "I'm not sure what time but I'll send you an SMS when I leave home". Another is flexibility in handling time "I'll send him an SMS to tell him that this meeting is running a bit late". The young people are taking SMS deep into social behaviour with dating including the preference that first dates be set by SMS rather than face-to-face or by voice call. Both boys and girls like SMS as it is less painful, is less intrusive, and both parties can consider exactly what to say – and when – rather than the often clumsy first words of initial contact. The reply to a date request can also be easily delayed for time to consider and perhaps even ask a friend's opinion "do you know this guy, is he a creep or a nice guy?" Some of the dating dimensions are explored more in the Me chapter.

Multimedia messaging

What is called multimedia messaging is various forms of added content beyond the basic text in an SMS text message. A simple analogy is the attachment to an e-mail message. The attachment today can be anything from a Word document, Excel spreadsheet or PowerPoint slide presentation, onto images and scanned photos, to music and video clips, and any other types of electronic documents and files.

Apart from the ability to send attachments, the mobile multimedia messaging has a few separate applications which are particularly suited for the mobile handset and the instantaneous need of communicating via messaging.

Picture postcard

The picture postcard is the first and most obvious use. This service is intended to replace printed postcards that people send when visiting holiday destinations and to replace greeting cards used for any instance of congratulating or celebrating or participating in another person's occasion or situation. So when visiting the Eiffel Tower in Paris, you could be offered a series of images that can be sent as postcards from your mobile phone directly to the mobile phone of any of your friends. The conveniences of not looking for a post card shop plus stamps plus mailbox, and that you don't need to type the addresses but rather use the phonebook on your mobile phone, make this a good substitute for the traditional printed postcard. The picture postcard via UMTS is much faster, being delivered within seconds or minutes, rather than the days it takes for a traditional postcard to arrive, and best of all, your friend can immediately reply by sending you a reply message, resulting in more communication as your friend participates in enjoying your trip. Of course the cost needs to be in line with that of the traditional postcard plus postage stamp.

Picture messaging

The picture message is another variation. As UMTS phones will increasingly have built-in snapshot cameras, it become very easy to take pictures from your trips and send them as multimedia messages to your friends and loved ones. Then, depending on preferences, you might take your own preferred view of the famous landmark, or take a picture of yourself at the famous tourist attraction to personalise the message. Very often we take snapshot pictures with an intention to share them with specific friends. Now we have to wait until the film is fully photographed, then take it to be processed for pictures, and then wait for the next time that the friend visits us to see the pictures. With picture messaging, the moment after we have taken the snapshot, we can already send it to the intended friends to see, and again, they have the ability to participate while the occasion of the snapshot is still happening. "Hi. This is me at the Monaco Grand Prix race".

Video messaging

Yet another, more advanced variation of multimedia messaging is Video messaging. Rather than sending a snapshot of what we want to share, we can also send a short duration video clip, such as the view from the hotel balcony, etc.

Unified messaging

As we have more and more means to send and receive messages on various platforms and devices, the need is arising for a single solution to the various messages. The Movement attribute brings additional needs and concerns, as ideally all of our messages should follow us in ways we want to where we want. Thus the unified messaging service needs a single queue for all message types. It has to have the ability to handle e-mail, e-mail plus attachments, voice, fax, pictures and images. The service needs to have multiple access including of course the UMTS handset, but also other phones and handsets as well as access from a PC, PDA, etc. Very importantly for the user, the unified messaging service would need a common user interface, with common features and the same way of operating it on all message types and all access methods.

Unified messaging would benefit from conversions between message types, including voice/video messages to e-mail with media attachments; text-to-speech for mail lists, headers and text content; playing compatible media attachments (sound to voice terminals); active notifications based on user defined rules, and notification via SMS or voice call.

There are a number of unified messaging products already existing on the market such as oneM@il from Singapore Telecom, a mobile portal which is called DoF (Department of the Future) from Swedish operator

Telia and Telecom Italia's Universal Number service. These need to be extended to support access from a wide variety of mobile terminal types including support for so-called 'thin clients' and the small screen size. Also needed is a common command set for all forms of messages.

4.7 *Organiser synchronisation*

Another closely related area is that of organiser synchronisation. As organisers we have various PDAs, PIMs as well as a wide variety of hybrid machines from very small sub-notebook PCs to the data organisers built into mobile phones. A whole new class of communication devices was born with the Nokia Communicator and its competing offerings from other cellular phone manufacturers.

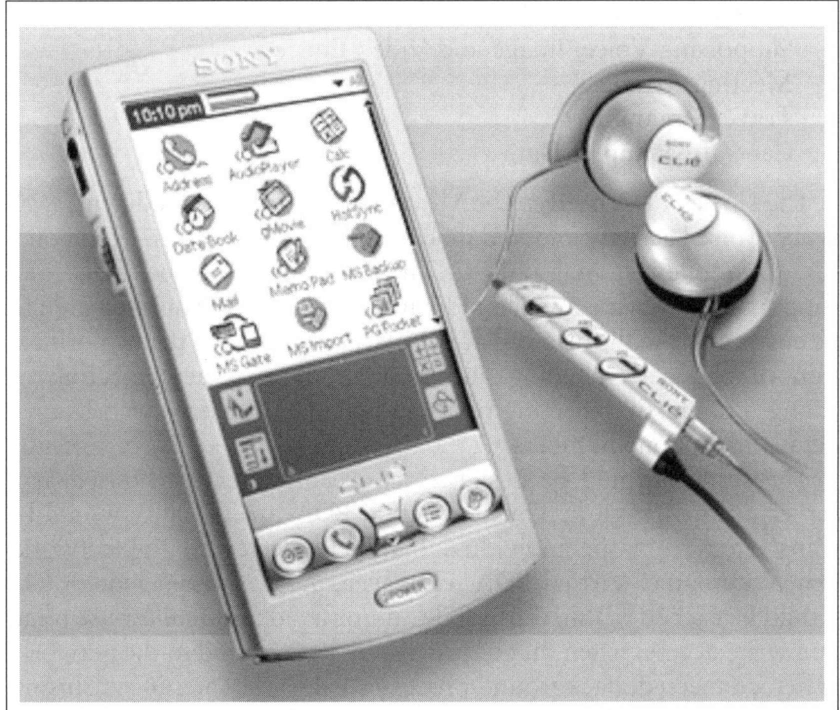

Personal Digital Assistants (PDA) are becoming more popular and are integrating browsing and telephony with traditional personal organisational features. The Sony Clié is one such example.

4.8 Virtual PDA

The current pocketable devices in the PDA and PIM categories are typically a collection of small utility applications designed to expand the calendar, calculator, database type that became possible with the advent of the microprocessor. Today most PDAs are used for purely personal use, much like first PCs before the advent of networking and the Internet. There is considerable added utility that can be derived from the Movement attribute, which much like networking did for PCs, will allow greater benefits. For the UMTS environment, a typical set of PDA applications could include the following:

- Diary Manager (Calendar)
- Task Manager & Action Chaser
- Contacts & Address Book
- Supporting Voice, Image and Video Entries
- Meeting Arranger
- Travel Organiser
- Central Backup of Data

The UMTS PDA needs to be Movement aware and intelligent, with synchronised activity and update between PDA versions residing on a server, mobile terminal and PC. The service needs to be secure and support both private and shared entries. The PDA should be able to be customised with rules and active alerts including dynamic notification of changes including automatic update of new meeting or changes to existing meeting times and change of users business card details like phone numbers.

The service would be designed as a central service housed at a server on the network. Thus the actual PDA functionality would be transferred from the individual handheld device to the network, hence the name, Virtual PDA. However, most of the personal data would be cached within the mobile and/or fixed terminal for use (read and write access) when that terminal is not connected to the network. When connected, background processes keep all copies in synchronisation, including the resolution of local updates with any network changes.

4.9 *Moving beyond movement*

This chapter has looked at some services that have a strong benefit on the Movement attribute of the 5 M's of Services for UMTS. The services were not an exhaustive listing, and are described mainly to provide a deeper understanding of the types of UMTS services that are likely to be created. Each of the described service would have several of the other 5 M's as a strong attribute as well and any of the services listed here could be mentioned in at least some of the other chapters as well.

Movement is the defining attribute that sets UMTS services apart from those on fixed networks. Movement has many sides to it and services will need to evolve according to customer preferences and be increasingly better at handling the Movement dimensions. Initially it is unlikely that a service could be designed to be perfect, so it is more important to experiment and be ready to adapt the services as customer preferences are uncovered. In designing UMTS services the operators will need to experiment, be creative even bold. Perhaps the Roman 2nd century BC playwright, Terence, can give us some guidance as he said: "Fortune favours the brave."

5

'I am having amnesia and déjà-vu at the same time; I think I have forgotten this before.'
Steven Wright

Services to Address Moment Needs:

Expanding the concept of time

Päivi Keskinen, Tomi T Ahonen and *Joe Barrett*

The ability to expand the concept of time includes being able to do multiple things at the same time – multitasking; to plan easily i.e. make good use of time; to postpone things i.e. make last-minute changes, or last-minute decisions; to do things with sudden idle time, such as waiting at an airport, i.e. to fill time; to catch up on missed issues, e.g. read yesterday's newspaper, etc., i.e. to catch up on time; and to get things in real time, such as stock quotes. Services that help control or postpone the moment will be of value to the user.

The services described in this chapter have a high benefit on the Moment attribute. There are hundreds of such services and we have a chance to explore only a few of them. But to illustrate by way of example some of the obvious uses of Moment type services, they include:

Access bank machine
Access and share information
Adult entertainment
Audio streaming
Audience feedback (vignette)
Business – market & competitor info
B2E (Business to Employee) services
Calculator
Cartoon services
Entertainment bundle
Film clips (micro-movies)
Fun dream analysis
Fun jokes
Fun proverbs, quotations
Games by download
Games puzzles & quizzes
Government notices
Info news
Info sports (vignette)
Info weather
Lending money
Local emergency services
Local taxi
Local time (for travellers)
m-Banking
Measurements converter
Music downloads
Music multicasting (vignette)
Music streaming
Opinion polls
Organiser agenda
Organiser calendar on server
Organiser calendar on terminal
Organiser reminder
Organiser to do lists
Promotion through desirable content (MTV)
Promotion with video clips
Sponsorship invisible to user
Sponsorship visible to user
Telehealth emergency care
Travel schedules

Moment is the most powerful activating catalyst of the 5 M's (Movement Moment Me Money Machines). When one has a sudden urge to gain access to a service or some information, it can be a very high value need and in extreme cases it may have almost limitless potential for its pricing. To give some illustrative examples, if you are on your way to an important business meeting and accidentally have your clothes soiled, for example by a passing car running through a puddle and splashing dirty water on you, then you have a sudden need to know if there is a clothing rental place, or a clothing store nearby. While normally you would not pay anything to find the location of the clothing rental store, or a clothing store, in this Momentary need you might be willing to pay very highly to satisfy the information need. If someone could catch you at that very precise moment, the person could perhaps sell you the 'worthless' information for several dollars, perhaps even up to 10% of the value of the new suit if your meeting was important enough.

Another example is if you are visiting an unfamiliar city or place. Let us assume that you know of one cash machine close to your office, but don't know of another nearby which would accept your credit cards. If in that situation you arrive at the cash machine and it is out of order and you must urgently find another cash machine, this sudden need could again have considerable value. If you are out of cash, must pay for your taxi fair by cash, and are in a hurry to get to the airport, again you might have Momentary value up to several dollars to know where is the nearest cash machine. Although we advocate that eventually the UMTS (Universal Mobile Telecommunications System) terminal will mostly replace cash in our society there is still going to be a need to find that elusive cash machine for some years to come.

We do not suggest that the operator could always hope to capture this type of exceptional need. But the momentary need can be a very powerful catalyst for action. The prudent UMTS operator will build services which serve the needs to manipulate time, to stretch time, to reschedule, to allow multitasking.

5.1 Mobile information

News, financial, sports and traffic information are types of push services that users can subscribe to based on need and interest. Some of these services may even be part of basic subscription packages the operators will offer to their subscribers as a way to differentiate their service offering. The value of these services is increased if the user is able to be kept relevantly informed – and up to date. For instance, users can receive information about the prices of shares and latest news about companies they have intentions of investing in. Pull services, like directory services will be enhanced with user requests based on location. For example, a traveller can request information via the UMTS terminal for the nearest petrol/gas station while driving in an unfamiliar district or country or request where is the cheapest petrol/gas within a 10 min drive. The directory can then send the addresses and location maps of the one or more nearest outlets.

Personalised messages can be sent to the user's terminal rather than distributed via mass communications. For instance, the user who is looking to buy an apartment can get updated information on available apartments in the district of their choice. They can then make an offer to the agent or landlord, or simply discard it, depending on the suitability or interest. This is a more direct way of getting relevant and updated information than through the newspapers or the Internet which is dependent on the user's active scanning and selection.

5.2 Mobile entertainment

The Moment need arises easily whenever the user has a moment of time to kill. This could happen at the bus stop, the airport, standing in line, waiting for a friend, sitting in a taxicab, almost anywhere we wait. These spare moments are opportunities to relax a bit and laugh a little with a joke or cartoon, or entertain oneself with a game, or catch up on the sports scores, or whatever brings entertainment value to the individual in a spare moment.

The entertainment industry has numerous aspects which transform well to the new UMTS environment. The music industry is looking forward to having music streaming and downloading directly to UMTS terminals which would also be MPEG (Motion Picture Experts Group) players. The gambling and betting industries are one of the most profitable parts of the fixed Internet and are looking to repeat their success when the gambler and betting person can have the opportunity for this service to follow them to any event and situation. National and cross-national lotteries are another area of similar interest. There are hundreds of industries related to entertainment which are all looking at entering the UMTS environment, from small jokes sites on the WWW (World-Wide Web) to the global entertainment empires such as Disney, Universal and Paramount.

Entertainment terminals like the Nokia 5510 will become more popular in UMTS as we see the combination of services into a single device. More categories of terminals will be developed that address these new markets.

Micro movies

The motion picture industry in Hollywood is working on 'micro movies' also known as 'micro cinema'. These are not motion picture advertisements or 'trailers' but rather complete stories told in 5 min, produced for the small screen format. For those who might doubt the viability of a 5 min story, one must remember that MTV (Music TV) **music videos** often tell stories and typically run for about 4 min in length. A short story can easily be filmed for a 5 min viewing. The micro movies need to be directed with more close-up scenes since an epic-like scene with thousands of civil war soldiers fighting it out as shown on the Hollywood big screen format, would not work on the UMTS small screen format. For the UMTS operator the best aspect of the micro movie is its pricing structure. The micro movies are priced to be free for the viewer, totally paid for by promotions that go with the 5 min film clip. So the promotions that can be built into the solution pays not only for the costs of producing the micro movie, but also for paying for the needed air time to stream the film clip. This is the format that works in TV all over the world so it is a relatively simple step to extend it into the world of UMTS. There is one clear difference though to the promotions delivered as part of the micro movie service when compared to the TV advertisement we see today.

The micro movie promotions will be relevant to the UMTS subscriber since they can be based on their user profile.

Entertainment bundle example

The services for UMTS will rely on techniques of bundling and segmentation. Typically a bundle offered for a given segment might have standard components and optional components for a set, relatively small fee. In this example it is assumed that the service bundle would be a basic simple entertainment package, with jokes, horoscopes, puzzles, games and cartoons. Typically the content would exist elsewhere and the owner of the content would join in the bundle. Various syndicates exist to provide aggregation services for the traditional media, and similar concepts probably will emerge for UMTS.

The user would get to select which services they want to have immediate access to on their portal viewed via the colour screen of the UMTS phone. Those services which are most attractive to the particular user would be the ones that the user would select. Typically the user will select 10–15 services that are relevant to their likes and lifestyle with probably a different set of services for the work profile and home profile. This could be 1–3 cartoon strips and for example only their own daily horoscope rather than seeing all 12. The service should allow the users easy access to other entertainment content, but not force the user to do any unnecessary scrolling or hunting for the favourite pages.

Each of the content providers in the bundle would have some kind of revenue sharing system with the operator, partially depending on the popularity of the service and partially depending on the network costs of delivering the content to the user. For a small text based joke, or even the simple graphics of a black-and-white cartoon strip, the actual load to the UMTS network would be very small.

As one example, let us assume that the basic entertainment bundle would cost 1 dollar per month, and that it could be made to be attractive to half of the subscribers. In a typical large European country, the UMTS operator might have 10 million subscribers and thus 5 million users of the basic entertainment bundle. The bundle would generate monthly revenues of 5 million dollars, or 60 million dollars annually for the UMTS operator. These revenues would need to be split with the content providers. Let us examine how that revenue might be split in the case of one of the content providers.

Note that the operator would have hundreds of such content providers in this system and each would get part of the revenue sharing. The overall billing of this service would generate 5 million dollars per month, or 60 million dollars per year, with minimal operator involvement beyond setting up the links to its billing system which is built to track billing worth cents anyway with voice minutes traffic. Most of the revenues could typically be returned to the owners or creators of the content but this is dependent on how each UMTS operator approached the content delivery business. It is however reasonable to expect the operator to retain, depending on service type, cost of delivery, the depth of information needed per user from the billing system and network loading anywhere from 10 to 50% per service. If we assume that the UMTS operator could keep on average 20% of the revenues from the services in this entertainment bundle, the operator's share of the revenues in this country would be 12 million dollars per year. Not bad for a 1 dollar per month bundled service and we are sure that many users would pay more per month for more content rich entertainment bundles.

Daily cartoon

Most popular cartoons like Dilbert, Peanuts or Garfield are currently available on daily newspapers and also on their free web sites. These typically show the daily cartoon, often with a little lag from the newspaper versions and store a few of the previous cartoons, such as all of last week or up to all for the last month. The web sites tend to be sponsored, so the cartoonist is likely to get some revenue from the fixed Internet site. Of course the cartoon authors get money from the newspaper version of their cartoon and also sell books with old strips and make more money that way.

In this example of the bundle costing 1 dollar per month, let us assume that several dozen cartoons are available, and each has a similar deal, getting only 1/10 of one cent from the monthly fee. With 5 million users of the service, even in this simple example, the cartoon service provider would receive 5000 dollars per month, or 60 000 dollars per year per operator per country! This as a wholly new revenue stream above all current revenues from a cartoon strip that the author and service provider are generating anyway. The key to success is to keep the enrolment cost below the threshold of the user. What is the most you will pay to laugh a little every morning?

It is very important to remember that every interested person who would visit that cartoon strip is a potential promoter of it. The service bundle should make it extremely easy for anybody to promote that site to anybody else by sending greetings from the bundle, offer possibilities for word-of-mouth, and even reward members to get more members into its service bundle.

Cartoon archive

The above mentioned cartoon example should be built to offer greater variety of pricing. The service provider should allow backwards access to unread cartoons since the cartoon fan would probably want to see any cartoons missed. The service should be built so that it keeps track of any which were missed allowing for easy access directly via hyperlink to these cartoons.

These do not have to be offered for free. The basic entertainment bundle should not allow viewing any archives Users would have to subscribe to the slightly higher cost bundle, say 1.2 dollars to view these and could be offered the upgrade if any attempted was made to access previous cartoons. Alternatively the user could pay a small single cost for yesterdays offering, a little more for all of last weeks and so on. Again the archive should be offered at the tiniest cost to encourage usage since familiarity breeds repeat business.

For example a good price could be 1 penny per cartoon. The user will most probably not mind this price since it is insignificant, but many might be put off by a price of say 10 or 20 cents. The key is to make the payment so tiny it does not deter ANY use of the service. If we assume that out of all who subscribe to the entertainment service, only one in every 100 is a real fan of a given cartoon, it still means there are 50 000 fans in that country, in the UMTS operator's network. If the archive service is priced at 1 penny per missed cartoon, and we assume that the fans would need to catch up on an average of 1 cartoon per week, there will be 50 000 archive requests per week at 1 penny per cartoon producing 500 dollars of extra

7 Vignettes from a 3G Future:

My Personal Portal on the Company Intranet

I had always felt it was hopeless trying to find information from the common data sources on the company Intranet. What made it worse was that when I had a sudden need for information, inevitably I was not connected to the company Intranet, and it would take forever to boot up the PC, establish remote connection and security, and log on. Now with the customized work portal for my UMTS terminal, I have constant connection so I don't have to wait to boot up, and access to those pages that are of specific interest to me, is only a click or two away . In my case I have tailored the personal portal to the work Intranet to include access to the company internal news, competitor information, certain marketing documents, and the menu to the cafeteria. The intelligent search robot even collects news for me and informs me of those new documents which are going to be of interest to me.

An operator can create a tailored Intranet portal access solution for businesses from very small to corporate. The smaller the business, the more likely that it would want to outsource the whole solution. The operator can tailor the services so that the Intranet access has various controls and safeguards to protect sensitive data. The system could integrate various work efficiency tools such as schedulers, mailing lists, internal database links, etc. A business personalised portal service would be particularly "sticky" in that it would tie into the whole way of working for a large portion of the employees for the operator, and thus ensure they would not want the service interrupted or moved.

income per week. Another way to look at the sums is to say this is 26 000 dollars per year of additional revenue on cartoons which have already been sold. This payment would not detract income from the initial entertainment bundle income either, as this service would be available to those who paid the 1 dollar per month for the basic package. For users who did not subscribe but were casual callers on the service the pricing could be 10, 20 or even 50 times that of the registered users without being price restrictive.

As with the bundle, so too with the archiving service, the UMTS provider and cartoon content provider should develop the tools to help promote the service and attract new interested parties. A logical use is to send greeting cards from this site, with a link to the site, as a promotion. The daily cartoon, or archived cartoon, should be able to be sent as a picture postcard with standard (or discounted) picture postcard fees that the UMTS operator has.

Cartoon fan club

More money can be made re-packaging the same content. Again the UMTS operator, the entertainment bundle manager and the cartoon content provider should entice the true fans to sign up onto a cartoon fan club. This should give clear advantages over anything on the fixed Internet site. So for example the fan club member could have free access to last years' worth of cartoons. The cost again should never be an issue that it causes the potential users to think too much about it. This service for example could be priced at 10 cents per month.

If we assume that out of the fans of the cartoon – 50 000 people, one in 5 is a big fan willing to pay 40 cents per month to join the fan club, it results in 10 000 paying fans. They would have full access to the cartoon content and they would not be billed separately for viewing missed cartoons. But with this simple extension the UMTS operator is generating an extra 48 000 dollars per year in revenue.

The major point to note here is that the possibilities of service differentiation with something so simple as a daily cartoon is awesome. When compared to the billions of dollars paid for the UMTS licenses in some countries, figures of 6000, 48 000, 26 000 and 5000 dollars per year sound like small drops in a huge ocean, but this is just one service. The total is in fact 139 000 dollars per year.

How many other small, inexpensive services will create even greater scope for differentiation?

If each of the 10 million UMTS subscribers has their own 5 entertainment bundles as described above then the total annual revenue becomes almost 1.2 million dollars. Now this sounds better. Total cost to the user? Between 5 and 8 dollars per month.

Selling single cartoons

Yet another way to re-sell the same cartoon to really die-hard fans and get further revenues, would be to sell a high resolution image of the cartoon for any given cartoon strip. Again, there is a huge pent-up demand to own any given cartoon, to really have the right to use it and have a high-quality image print of it. Printouts from the Internet versions, or enlarged photocopies from newspaper versions of cartoons do find themselves on various office walls, but are hardly the stuff to frame and put on a wall. The UMTS operator or service provider should feed this need as well. If this threshold is again kept really low, the uptake of the service would generate a lot of revenue. We can expect that a cost of 1 dollar is a reasonable cost for a high quality image, i.e. not the right to reprint or re-sell it. If this purchasing process is made very easy and the billing is taken care of via the UMTS operator's billing and micro-payments system, then a huge potential market is opened.

Now all of the die-hard fans, who pay 40 cents per month to be part of the cartoon fan club, can also own their own favourite cartoons. In fact they could even send gifts to friends or colleagues. At 1 dollar per cartoon, it would be fair to assume that die-hard fans on average might purchase 2 cartoons per year. At 10 000 paying fans in the fan club, and at 2 cartoons per year at 1 dollar each, the Operator re-sells the same content to the same audience, maybe even for a third time and generates another 20 000 dollars per year. Another small drop into a swelling ocean.

Public use licensing of digital content

The cartoon owner might even want to make it easy for businesses to own rights to use the cartoon but not to sell it via the UMTS service. A use might be for example as illustration material in a public presentation at a conference. The right would be very specific that the cartoon may be shown but not sold, nor reprinted, including not publishing in handouts of materials, etc. The price of such right to show but not reprint the cartoon, could be priced at 10 dollars per cartoon and since the take up would often be corporate users the price of this service is more elastic. The cost is still well below that of handling a separate credit card payment and would serve a large pent-up demand. Many of the real fans of the cartoon would like to use it in some context at work. If we assume that only 1 in 10 of the paying fans would also have a business use for a slide, and assuming that the fan would only purchase one such cartoon per year, we are again looking at another 10 000 dollars per year.

Of course the prudent cartoonist will provide here the link to his legal advisers for those who want to actually print a cartoon in a corporate brochure for example, to provide access to even more revenue from the same content. But that payment would not qualify as a small payment and is not worth examining in this context.

The total revenues for the operator, out of this Entertainment bundle would be 60 000 dollars, and additional revenue streams could be built on that customer base, worth 42 800 dollars annually. All of this income would be from one UMTS operator in a large European country. The same model can be duplicated across all UMTS operators. The revenues would be generated from content which is currently produced for fixed Internet use and is currently offered for free. Adaptation cost are then relatively low.

The additional cartoon traffic would create further load onto the network and require revenue sharing, but the overall value of the service would generate over 100 000 dollars of revenues from the cartoon and its fans, to be split with the content provider, UMTS operator and any other related parties. The content producer gains extra revenues from work already sold. The operator gets addictive content onto the UMTS network which will generate more traffic and revenues. And the consumer would enjoy multiple ways to access content that is actually wanted. It is a classic win-win-win situation and clearly a way to make money in the Mobile Internet, out of content which is currently provided for free on the fixed Internet. All that is needed is a mobile application to make it happen.

This is one of examples of the magic of small payments, large populations and the business logic of micro payments.

5.3 *Music*

There are few people in the UMTS business who would deny that downloading music to the MP3–MP4 enabled UMTS terminal will be big business. What is uncertain at the moment is exactly how the music industry is going to be able to control access to its royalty bearing content and what is the business model in the future that will allow them to still make money? Music will eventually go digital and it will be delivered to mobile terminals over the UMTS network to 100's of millions of users. This will happen because people want it to happen and in the Mobile Internet that is what will drive change.

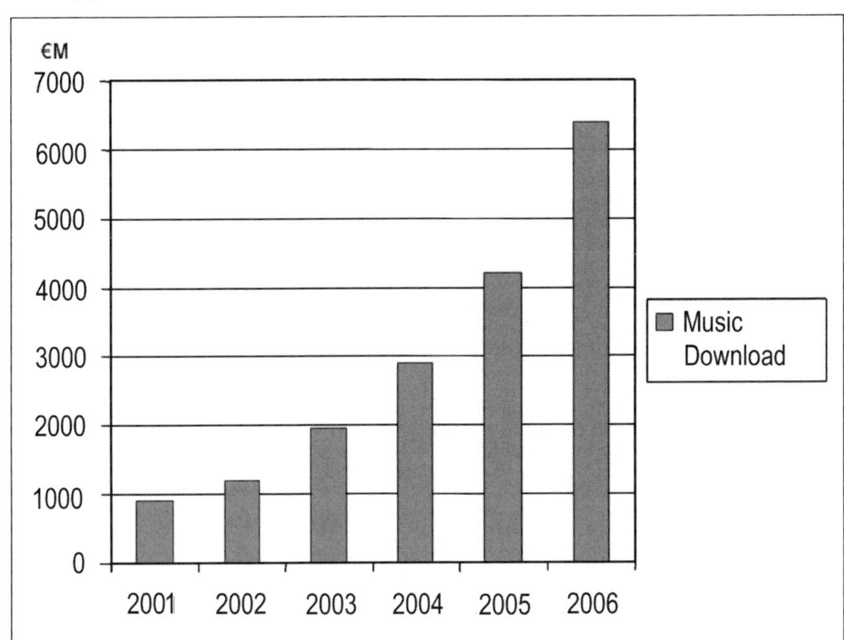

Source: IDATE 2001.
Music will be one of the major categories in the entertainment segment and will gradually capture a growing percentage of the total music sales market. Downloading music to new entertainment type UMTS devices will expand as the cost of data memory prices fall.

Mobile music delivered to the MP3–MP4 enabled UMTS terminal has clear advantages over the traditional music business model. The most important advantage is that a huge part of the distribution chain has simply vanished and along with it a large portion of the costs associated with getting each particular song to individual consumers. Although the music industry is fearful of losing control over its distribution and royalty it also needs to fully appreciate the opportunities that are offered by the UMTS enabled Mobile Internet.

This change is already happening. Sprint PCS (Personal Communication Systems) in the USA announced in November 2000 that it was to launch a new MP3 on line music service together with

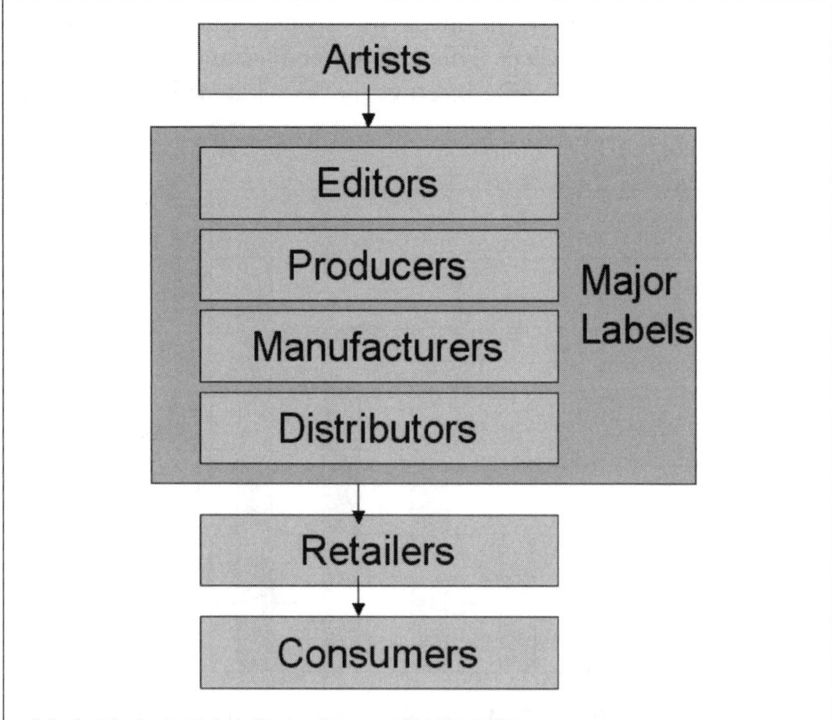

Music Market's Value Chain. Source: IDATE 2001.
As more music goes online and is downloaded electronically as ''bits'' of data the existing Music Value Chain will evolve and become more efficient and streamlined. The role of retailers and distributors will change and some artists will develop direct channels to their fans. This will only happen however when Digital Rights Management that protects Artists' copyright is introduced.

Samsung who would provide the MP3 enabled mobile phone. Users of the service would get one year's free subscription to the 'My Music' service when purchasing the phone. Digital music download to the mobile phone has already started and is here to stay.

It is important to always remember one clear difference between the Mobile Internet and the fixed Internet. The latter is based on the 'free lunch' business model where surfers have become accustomed to accessing content and downloading data for free. The Mobile Internet on the other hand is a fully paid up member of society where users 'expect' to pay for a service that they value. Once the fear of the Mobile Internet is dispensed with it is possible to look at the opportunity that exists to deliver music to the billions of mobile devices that will be in the market in the not too distant future.

Mobile music fan club

Few of the music companies have caught on to the promotional aspects of the fixed Internet. Although there are many artists who have their own web pages few of them use these pages to effectively promote their music or their tours. Often the clubs that provide the information that fans need are run by other fans and as such are out of the control of the music publishers and artiste. In the web it is content aggregators like Netscape, Yahoo and AOL (American Online) that are hosting Music sites and other Internet fan clubs mainly deliver static content and information and few if any are sending information and news via even a simple SMS (Short Message Service) service. Some will send e-mail news if you give them your e-mail address. The problem is that then you get onto a music e-mail list and we all know how the story goes from here.

It is clear that there is a need for a more focused and organised music fan club solution that will actually serve the fans that often idolise these members of society. The value of being able to create a more personal relationship with fans has to be profitable even in the short term for the music industry. Delivering short news about what the band or singer is doing, when the new single or CD (Compact Disk) is available, giving prior notice of tour dates even by SMS has to be good for the business. The revenue analysis used for the cartoon service can easily be applied to Mobile Music fan clubs. Many fans would subscribe to this kind of service if the information was delivered to the UMTS terminal. Even better if the music could be downloaded as well.

Mobile music download

The advantages that UMTS bring to music downloading are firstly that since the mobile terminal is the one device that is always with us, it is also the ideal device for storing and listening to music. Secondly, the distribution chain is far shorter and music can be delivered at a far less cost with increased profitability. Thirdly, targeted promotions can be directed at real music users who conform to the profile of different bands, singers, music types and styles. It would be a relatively simple matter for entertainment promoters to diversify and extend their product offering in a similar way that the film industry has done with merchandising. Top Hollywood stars now ensure they get a percentage of the merchandising profits since these profits can gross as much as the movie.

Music download also brings a new dimension to revenue creation. In the UMTS network although the user pays for the music that is downloaded to the mobile device, that is all that is paid for. As an example of this a music fan may receive a video message (Multimedia Message) at some moment in the day when they are not busy, 'sent' from their favourite band which they download and view. The option is given to purchase the video song and downloaded it directly to the mobile terminal or to purchase the video CD using secure mCommerce and have it delivered to home. No matter which option is chosen it is only the cost of the music, the single video song or video CD that is paid by the music fan. The cost of delivering the video message (promotion) is paid for out of the music company's marketing budget. The cost of downloading the data (video song) to the UMTS terminal will be part of the distribution costs paid for by the UMTS operator. This replaces the costs of copying music CDs, packaging them in boxes, shipping to the music store, etc. Now the cost is digital distribution via the UMTS network. That distribution cost will be part of the overall cost of purchasing the music, and no separate fee will be charged for the download traffic. Of course in either case, the user sees the music on the phone bill.

Now the operator has two new revenue streams. Firstly from the music company who are promoting the music for sale and secondly a substitution revenue stream from the music store who would have made the sale if they had been able to reach the music fan.

Audio streaming

A mobile phone like a radio? Audio streaming is already a reality on the fixed Internet and will simply slide across to the Mobile Internet with some clear enhancements. Mobility is one aspect but it is always nice to find that there is a spare moment and then have the ability to listen to some music. Additionally to this, while the user is listening to the streaming music it is simple and easy to purchase the current song and download it to the mobile device. This is clearly a strong moment aspect of the Audio Streaming service since it can generate impulse purchases, even more so if the relative cost of the download is low when compared to the perceived or realised end user value.

Audio streaming will also be an ideal service for promotions and adverts, even sponsorships. Once again it should be ensured that the promotions and even sponsorships are based on the users profile.

5.4 Mobile banking

Mobile banking data services have been in commercial existence since 1997, when the first banks and operators in the world started implementing SMS-based mobile banking services. At the end of 1999 there were already more than 30 mobile banking (SMS and WAP (Wireless Application Protocol)) services in place world-wide[1]. For banks and financial institutions, the mobile internet extends their reach to customers, while offering operational cost-saving benefits. The mobile banking services have also often been a natural next step from the banks' fixed internet or (voice) phone services. From the mobile operator's point of view, banking services provide opportunities to increase mobile traffic. End-users get the benefit of being able to handle their banking whenever and wherever they are.

[1] Durlacher Research 1999, p. 41.

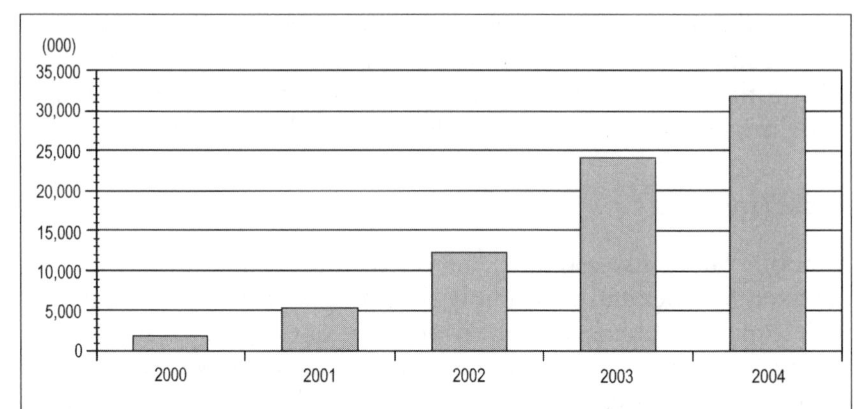

Number of Western Europe WAP-Enabled mBanking Accounts by 2004. (2000-2004 CAGR is 104.7%). Source: IDC 2001.
The mBanking capable mobile phone will create a huge leap towards the cashless society that economists and bankers have been dreaming about for decades. Whether it is with WAP or other supporting technologies the growth is expected to happen quite quickly and may yet become one of the ''killer applications'' of the UMTS business.

Security concerns and technological limitations mean that many mobile banking services are still limited to providing account information via SMS or WAP. With UMTS services these concerns will be overcome, as security and authentication standards will be developed, and mobile banking will gain even greater market acceptance[2]. PKI (Public Key Infrastructure) – an asymmetrical encryption method using digital certificates – and the use of digital signatures are widely considered to provide the requisite level of security for mCommerce. Digital signatures can be implemented by using a smart card (SIM (Subscriber Identity Module) or a separate card) as the cryptographic token.

Standard mBanking services

Despite the security concerns, there are already some nearly full-service mobile banking services available, such as Merita Bank's

[2] The Yankee Group 1999, p. 2.

WAP-based service in Finland. All customers of the bank can currently get bank account balance information, pay bills, buy and sell stocks and other financial investments, and shop with their mobile phones. Customers use the same user ID and (randomly generated) passwords as with the bank's phone and Internet interface for the banking service. Merita Bank has its own WAP gateway and the users dial-up directly to the bank. In other words, in this case the access to the mobile banking service is operator-independent.

Possible new mBanking services

As with any services for UMTS, it is impossible to give an exhaustive mBanking list. Some of the mBanking services that may seem more exotic today are still simple UMTS services in that they tend to make time and place disappear, while empowering machines and extending one's self regarding money (fulfilling all 5 of the 5 M's – Movement, Moment, Me, Money and Machines).

Mobile cash from cash machine

One possible service is to access a cash machine and get cash. There is nothing strange about putting in a banking card and accessing cash from the cash machine. There is nothing to prevent designing a UMTS service which allows access via the UMTS phone and one's own bank account, to get the money from the machine.

Lending money to friend

Another need that often exists is the need to lend money to a friend or relative. Again if it is possible to send an SMS message to the friend and it is possible for both parties to do bank transactions via mobile phone access to the banking service, then there is nothing technically to prevent direct lending to a friend – in effect making true self-service universal banking possible. It would be much like lending paper money directly from the wallet.

Paying bills

With UMTS many of the banking services can be automated. Your electricity bill will be sent to your UMTS phone for payment authorisation giving the date of when it should be paid. An alternative payment date could also be entered. A simple 'OK' accompanied by a PIN (Personal Identity Number) will complete the authorisation which should have taken a matter of seconds. A click-to-talk service request would give direct access to a customer support person at the electricity company if there was a query about the bill. For the customer this kind of transaction should be free or part of a low monthly subscription. It would be ideal if this kind of service was always part of the standard UMTS service bundle. The electricity companies would save millions of dollars in administration and postage cost and could be notified instantly when each customer has authorised their electricity bill to be paid. This will give early visibility of when payments will be received. Operators would be able to charge for the delivery of payment invoices thus creating a new revenue stream. In this case there is a three way co-operation needed between the bank, the electricity company and the operator. Each has to play its part in the overall process to make the service seamless and easy to use for the mBanking subscriber.

In most Internet markets, eCommerce has been preceded by the establishment of eBanking[3]. The same development is now going on in the wireless world. The UMTS Forum predicts that in 2004, there will be nearly 250 million mobile banking users in the world[4]. Although GPRS and UMTS technologies may not greatly enhance the mobile banking service content of today (although they will certainly increase the usability), the adoption of mobile banking by the end-users will pave the way for many other applications, especially mCommerce and micro-payments (or 'mobile wallet'). End-users may also justify their Mobile Internet/UMTS terminal purchase decision with the ability to use the 'serious' applications such as mobile banking or stock quote information, but in reality they might much more frequently use entertainment services. This has also been a widely quoted phenomenon with the Japanese i-mode service.

[3] The Yankee Group 1999, p. 15.
[4] UMTS Forum 2000, p. 11.

Roles of players in mBanking

Although many mobile banking services were originally implemented as a joint effort between a bank and mobile operator (e.g. Barclaycard's and the mm0_2 service in the UK), it can be assumed that in the future we will see some arm-twisting over the service and customer control between the banks and operators. The banks are likely to want to be in control of the access to their own services (like Merita bank in Finland); the main reasons being security and the reach of all the bank's customers, who may have different operators' mobile subscription. This is, however, not the only option. Especially smaller banks may find it useful to be placed on an operator's portal or even co-brand services with the mobile operator, and let the operator handle the hosting of the mobile data access and security solutions, thus being able to concentrate on their own core business. For example Sonofon in Denmark has been providing Danish banks with this type of commercial concept. In a successful partnership scheme operators may even expect revenue sharing with the banks, who are able to charge their customers a premium for mobile banking access.

Currently there is clearly some overlap and even competition between the operators' who are getting closer to the customer and are looking at integrating their business model into as many businesses as possible to recover their UMTS license and network investments and the banks who have in many cases an old established image that over the years has become tarnished in some countries. Banks need to move into the 21st century and the Mobile Internet and mBanking is one way they can do this. What still has to play out is how any co-operation between the banks and the UMTS operators is going to be eventually formulated. We have already seen that UMTS operators' with their ability to service the micro-payment market and the anticipated rise in mCommerce transactions moving to the mobile phone bill could almost become banks in their own right.

Some operators are already considering how this change in behaviour that they can drive with UMTS service will effect their strategy and if, when and how they might offer more mBanking services in direct competition to the traditional banks and the growing number of Internet Banks. What is most likely is that there will eventually be some form of partnering and sharing of competence that will bring

8 Vignettes from a 3G Future

On-line Card Games

The plane is delayed and my four companions and I are stuck for an extra hour at the airport. One of my colleagues suggests that we play a couple of rounds of poker to kill the time. We don't have playing cards, of course, but we all have our UMTS phones. The friend knows of a UMTS service where teams of people can play for free. I ask how come it is free, he says it is sponsored by a big on-line casino. We take out our phones and play, and our time flies. And as I am quite clumsy with my hands, I am happy I don't have to shuffle the cards.

Games are expected to be one of the biggest groups of new services on UMTS networks. Hundreds of game developers have already developed games for small portable devices, and tens of thousands of games already exist for the mobile phone. With games the content is likely to be sponsored or separately billable depending on the amount of load that the game would place on the network. Many business models for games include first demos for free and payment of full games or subsequent games.

about a win-win situation. For this to happen the banks have to recognise that they do not have the skills or knowledge or market position to offer the most advanced and user-friendly mBanking services without the UMTS operators. Likewise the UMTS operators do not know the banking world, which as a whole is about much more than arranging simple money transactions. Banks provide a wide range of financial services and if they get it right with the right partners then we will see extensive take up of mBanking over the next few years.

5.5 Mobile games

Mobile games could be such a big and revolutionary business that a whole book should be written just about mobile games. The electronic gaming industry features names such as Nintendo, Sony Playstation, Sega and Electronic Arts. The industry is the fastest growing segment of the entertainment industry by a wide margin and in volume of sales has reached the size of the motion picture industry box office revenues. It is reported that the games industry had reached 20 billion dollars world-wide annually, growing at annual rates of about 17.5% which is the fastest growth in the entertainment industry by a wide margin[5].

The games exist in incredible variety, from racing car games to role-playing games; from space games to the now famous i-mode fishing game; from animals, heroes, popular characters to shoot-em-ups. The games are geared towards the young and youth-oriented segments and thus they play on fantasies and allow a lot of creativity and imagination.

The current games are typically delivered on the portable handheld gaming devices, on devices that play on TV, and on PC based solutions. There are some primitive – but addictive – games that can be played on the mobile phone, but the low screen resolution has limited the use of early mobile phones as a gaming platform. The gaming industry has pioneered revolutionary bundling, cross-advertising and sponsoring methods, including putting games on other devices where they did not exist before, such as the Snake game in Nokia phones and the

[5] The Economist, E-Entertainment Survey, October 7, 2000, p. 10.

various game-oriented motion pictures such as Super Mario Brothers which was designed to promote the game and generate more revenues from the game of the same name. With PC based games the industry learned to take advantage of the 'first game free' model where subsequent games with different, levels or difficulty are then sold. The industry is driven by young eager and devoted programmers who want to create the ultimate games. The industry is known for its innovation and also its very fast reaction ability to change and opportunity.

Those developers have been looking for more resolution on mobile phone screens, more portability on the terminals, more interactivity and ideally connectivity to support multi-player remote gaming possibilities. The gaming industry is also very shrewd in understanding the money side of the business and very many gaming companies are very profitable. In UMTS they have a new platform which combines all the necessary elements and the gaming industry is primed and ready to do business. In effect, UMTS provides the perfect environment for games.

The UMTS terminal is the ideal gaming platform. It has a relatively high resolution screen, better than a GameBoy for example which are very popular gaming machines. The terminal itself is pocketable and small. The terminal will reach penetrations that are much larger than the whole PC penetration within a few years and the UMTS network will deliver a lot of capacity and allow game updates, remote gaming and multi-player gaming. Additionally the UMTS terminal will be always on, so interactive games can have real-time updates to scores for example. Getting a message that your Tetris score has been beaten by friends will encourage competition amongst gamers resulting in more games being played.

Games are found to be particularly addictive to their users. The most fanatic gamers do not care about the time, place nor cost. For the UMTS operator, game players are an ideal target audience. They are already accustomed to similar devices – gaming consoles tend to have small screens, small control buttons, and various menu driven instructions. Game players have yearned for multi-player options, to play with friends together, or to play with friends or strangers remotely. As some children – boys and girls – in Finland already say when parents ask what is it they need to pack on a trip to visit the relatives, the child will say the first two things to take along are the mobile phone (number one) and the Playstation/Nintendo/equivalent (number two).

If the people born in the 60s and 70s are the TV and MTV generations, the people born in the 80s and 90s are definitely the gaming generation. This generation is now in high school, college, university and starting in their first jobs. This generation is perfectly poised for the game developer wanting to expand its business. In UMTS the game developer has a platform where all users are known and all users can be billed. UMTS is the ideal environment. It provides all the components to create fun and make money.

5.6 Adult entertainment

One should not talk about entertainment for a new media, without mentioning one of the biggest and most profitable parts of entertainment in *any of the other media world-wide*. The adult entertainment industry is a very profitable part of most media including magazines, movies, video rentals, etc. The adult entertainment services are probably the *most lucrative* part of the fixed Internet today and early on they were probably the only industry who could make money out of the fixed Internet – although since this industry is very secretive, precise data is not readily available. The adult entertainment industry has pioneered automatic links, free but censored previews, various payment methods, age screening, etc. In the fixed Internet many of the service innovations found in other site categories actually came from early adult entertainment site examples.

It should be noted that the adult entertainment industry is one of the very few industries which has been able to generate *value-add services to fixed telephony*. The business in fixed telephone premium cost call services is so good, that in all markets where they are legal, adult services advertise their phone numbers on local TV at night. The adult entertainment industry is looking at how they can exploit the opportunities of high resolution video screens and the immediate connectivity aspects of the UMTS environment to deploy their services. The adult entertainment industry is likely to be one of the early profit areas of UMTS depending of course on local regulations and to the degree the industry is active in a given country. It is to be anticipated that any adult entertainment service that is currently available over existing fixed telecommunications networks, either voice or data will be available on UMTS networks, based on premium rate services and will be available from network launch.

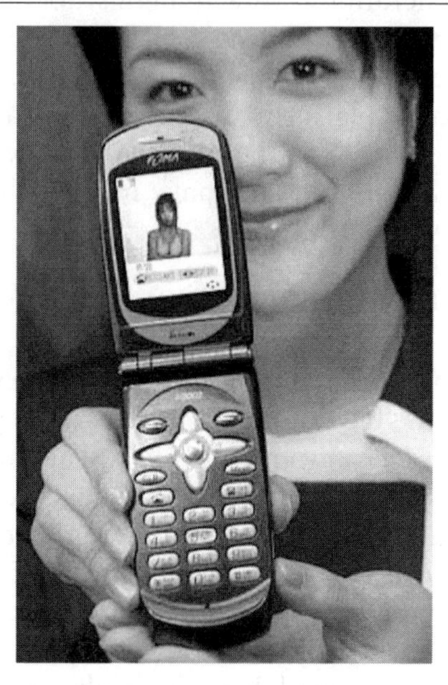

Video phones will become more and more popular with UMTS offering a range of services. The miniaturisation of camera technology is driving the market towards digital imagery that is part and parcel of the mobile phone package. In the not too distant future all UMTS phones can be expected to include a digital camera for either still or moving pictures. This example is of an existing NTT DoCoMo 3G FOMA video phone from NEC.

5.7 Last moment on moment

This chapter has examined the Moment aspect of services for UMTS. The other aspects of the 5 M's are of course important as well, and none of the mentioned services relies only on the Moment attribute. Trying to use time, save time, move time (reschedule), multitask, and manipulate time in any way we can is an ever increasing need by humans. Services which help in doing so will bring value to the user. The future will be here soon and in that future, man or woman will be ever more stressed and concerned about time. Perhaps all is not lost, however. For the UMTS operator it may be comforting to remember that nobody knows what the future really holds. To quote Scott Adams in his book: "In the Dilbert Future it will become increasingly obvious that your competitors are as clueless as you are."

6

'I don't care to belong to a club that accepts people like me as members.'
Groucho Marx

Services to Address the 'ME' Needs:

Extending ME and MY Community

Russell Anderson, Mika Suomela and *Joe Barrett*

The Mobile Internet's ability to provide personalised content and to extend the self into the near community will be highly relevant in UMTS (Universal Mobile Telecommunications System). This includes personalised content, exclusively relevant content and custom made and modifiable content i.e. the ability to interconnect and share with the community, which could be family, work colleagues, friends, clubs, etc., to communicate in one's own language, and to allow multi-session use, i.e. view a promotion for a movie while also talking on the phone about the movie.

The services described in this chapter have a high benefit on the Me attribute. There are hundreds of such services and we have the chance to explore only a few of them. To illustrate by way of example some of the obvious uses of Me type services, they include:

Astrology	On-line card games
Business – CRM (Customer Relationship Management)	Organiser community calendar
Caller picture presentation	Personal assistant call management
Chat	Personal assistant text-to-voice
Cinema	Personal assistant voice-to-text (vignette)
Community calendar	Photo albums
Company internal phone book	Picture postcard
Dating	Private group bulleting boards (e.g. family, corporate)
Entertainment for adults	Private personal greetings (vignette)
Games as real time connection	Promotional permission pull
Horoscopes	Public bulletin boards
Info culture	Rich call
Jokes	See what I mean
Local directory assistance	Translation of text
Local TV	Translation voice
Mobile portal	Travel frequent flier programs
Multimedia messaging	Video call
Multi-player games	Video conference
Night life guide	

Among the services which we feel are most personal there is our own voice on our own calls. One could say that the ability for us to call someone else on our private mobile phone is the ultimate service using the Me attribute. Voice calls themselves, however, are nothing new. Bringing voice calls to the mobile handset created the more personal dimension to the phone. Many of the older generations will remember that the mobile phone concept was actually popularised by secret agent Maxwell Smart on the 1960s television comedy show Get Smart. Maxwell Smart had a mobile phone in his shoe and always removed the shoe from his foot to make a call. Now with UMTS there will be radically new tricks to our favourite old telephone call, called Rich Calls.

6.1 *Rich calls*

Voice telephony will without doubt remain a very important application category in the future and on UMTS. Telephony itself, as part of

IP multimedia, will be enhanced through the possibility of allowing the user to not only 'listen to what I say' but also to 'see what I mean'. Rich calls, that is audio conversations supported with concurrent access to an image or data, will greatly enhance personal communication. Unlike fixed video telephony, mobile video telephony provides flexibly real user value by allowing people to see what is being discussed, not only who is discussing.

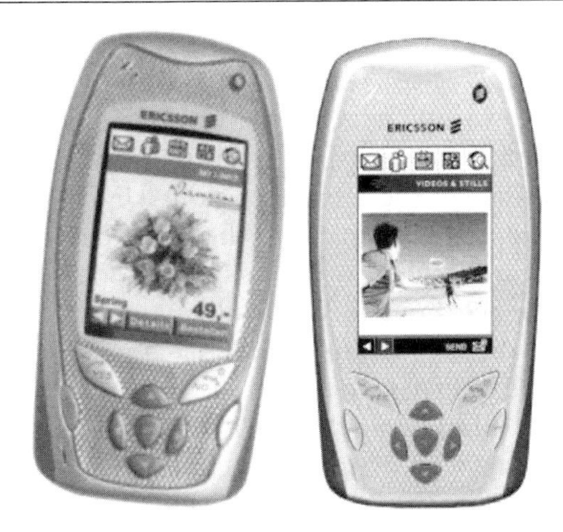

Imaging will play a large part of UMTS phones and the majority of terminals are expected to eventually incorporate large full colour screens and cameras.

Draw it for Me (or Whiteboard)

A useful example will illustrate what can be done with Rich Calls. If you are on your way to your good friend's new home, following his directions, but get lost along the way, it would be nice if you could have him draw a map for you. This can easily be done with UMTS with the Draw it for Me service, also known as Whiteboard. As you call your friend to say you are lost, and are at the Shell station on Smith Street, your friend can draw the map on his UMTS phone, and you will see it on yours, simultaneously as you speak. He and you can both now discuss the exact directions, where you are and where you are to go.

Draw it for Me is only one of numerous Rich Calls, where you add images, moving pictures, games, or chat to your live phone call. These kind of rich calls, calls allowing concurrent use of several media types, let users access and share information while talking without the need to get the information first and then call back. By introducing rich calls, users will eventually become accustomed to multimedia telephony and accordingly raise their expectations until rich calls become the norm.

Share picture

Another Rich Call service is sharing a picture or image. Just like in Draw it for Me, we might be in a live voice conversation on our mobile phone, when an occasion arises to want to show a picture. If that image resides in the memory of our UMTS phone, it will be easy to open it up and transmit it to the other person we are talking with so that we can both enjoy the image. A good example would be the new baby in the family. We could have a few images of the newborn baby taken with our snapshot camera feature built into the UMTS phone. Then when we call up our various relatives to discuss the happy family event, we can not only talk about the beauty of the baby, we can send images of the baby during the call and the other end can see the same pictures as well.

Voice portals

As voice processing capabilities mature in the coming years, voice portals will take ease of use into a new level. For instance, speech recognition and processing enable voice browsing of internet based content such as airline timetables and ticketing. This means that keyboard size is no longer an issue in mobile devices, as most commands can be carried out by speech which is the most natural of all input devices. The user interface can be single mode where all responds are converted into voice, or multi-modal where there is a combination of verbal commands and graphical actions.

Rich Call type of services can be provided in the first phase of UMTS launch with new service platforms and functions. This allows operators to create and test new services and provide customers with something new and truly exciting.

Voice related services can be divided into three categories:

- Basic telephony services provided by CPS (Call Processing Server)
- Rich Call type of supplementary services provided by APPSE (APPlication SErvers), utilising the CSCF (Call State Control Function) of CPS
- Virtual presence type of services provided by a 'Community Server' not needing any call state control

On the basic level, virtual presence means that a user is logged into one or more user groups in an application server (professional, family or hobbies related groups) stating the presence status. When active, for instance in a football club group, members can send instant messages, chat with text or voice, or start a voice conversation controlled by the application server.

6.2 WTA (Wireless Telephony Application)

The WTA (Wireless Telephony Application) is the first stepping stone towards IP (Interent Protocol) telephony rich calls. WTA acts as a 'missing link' between WAP (Wireless Application Protocol) content and telephony related functionality of network and mobile device. WTA provides a user interface and signalling mechanism for speech services, enabling rich call type of functionality. The user can detect and react to network events like allowing the user to receive, deflect or transfer the incoming calls in real time, instead of pre-set transfers or used to initiate calls such as 'click to talk' from yellow pages.

WAP can be used to access and control most any kind of service or application – it may or may not be telecommunication related. WTA is part of the WAP specifications where the nature of WTA is telecommunications and messaging related. The role of WTA and the WTA server however is not limited only to the WAP front-end for a telecommunications network. The WTA server can be used for various applications and application areas.

WTA service examples

A few examples can illustrate what WTA can do. **Personal WAP Homepages** can be created much like personal WWW (World-Wide Web) homepages on the fixed internet. The user can customise their mobile home page or portal for their own specific requirements. **Charging Services** can be created allowing user interaction and control of different payment methods and tariffs. The user could for example define when and where they could receive specific services or messages so that they are delivered or used at times when the service is priced according to their needs. **Call Management Services** allow enhanced call control applications and user interfaces. UMTS will enable more user control of their services putting the power more in the customers hands. WTA also includes location-enabled abilities and personal profiling services. Location and its related Movement services were discussed in the Movement chapter. Profiling will be discussed in more detail later in this chapter.

6.3 Video calls

Video telephony has been the great hype in UMTS just as it was for a short period of time in the fixed networks. As long as we can remember science fiction movies and television series have shown that in the future all phone calls will be video calls. Comic book heroes such as Dick Tracy with his wrist watch phone, and the Jetsons and other visions of the future invariably included video calls of some kind or another. Apart from strict Video Telephony, there are some 'milder varieties' of calls with moving images, such as Rich Calls discussed before.

Although video telephony is already fully available on the fixed telecommunications networks it is still quite seldom used and only the tiniest fraction of worldwide telecommunications traffic is video calls. Video telephony is currently almost solely used for corporate video conferencing. Business users are becoming gradually more accustomed to video conferencing as a way to reduce travelling and save costs but very few consumers have even tried a video call. Video telephony will continue to be the main hyped service towards consumers since it sounds exciting and new.

Mobile Video will be one key UMTS service and a number of products are already being developed with this in mind including this fully integrated video phone from Origami.

The sci-fi future of video telephony is possibly just around the corner for most users and this sounds so much sexier than the simple low bandwidth services that they have become used to. If you tell future UMTS users that they will mostly make low bandwidth multimedia calls and that entertainment will likely be the big service they may take a less revolutionary view of what UMTS will give them. However the evidence is there that people do want to use video type services and this includes video telephony. We just have to make it happen.

Is there great potential in video telephony?

Estimating the popularity of video telephony in the future is the most difficult part in UMTS service analysis. The expert opinion and early user data seems to suggest a rather robust consensus on most of the categories of UMTS services. We can be pretty sure that people will use mCommerce and that they will consume mobile entertainment services, send multimedia messages and they will use their mobile terminal for e-mail. When the ability to use video calls will exist, some will become users. The question is, how much?

Change can come rapidly

Nowadays we wonder why anybody would use a fixed telephones anymore. In many European and some Asian countries already today the under 25s are more likely to not have a fixed line for voice using only their GSM (Global System for Mobile communications) phone for all calls. A call on the fixed phone can be irritating for some people since it means that you cannot walk to the coffee-machine at the same time as talking on the phone or find out why the children are screaming at each other in the bedroom. Even more dramatic is the rapid adoption of SMS text messaging. First SMS (Short Message Service) messages ability only emerged in the mid 1990s and only five years later, the SMS traffic generates 10% of operator revenues in most Western European countries.

After five or ten years we may wonder how people could ever have used phones without seeing each other. On the down side this could be irritating for some people since it means that the caller can see that you are walking to the coffee-machine at the same time as talking on the phone or that you have very badly behaved children. The option to have the video function activated only when you want will have to be an option. Personally we would not like anyone to see us first thing in the morning. And there are occasions when executive don't want the other caller to see exactly where they are either.

6.4 Show Me

Video telephony does not just mean person to person, it can be that the caller shows the environment or what it is the caller wants to ask about. For example an anxious husband worrying about what to buy for his wife for her birthday so he shows her what he intends to purchase, or even get the remote advice of her sister or best friend. There will inevitably be the type of controlling personalities who might send their husband shopping to the supermarket and the wife would then give instructions via the video phone on what meat, vegetables and other goods he should purchase.

There will be occasions while travelling on a business trip say to Hong Kong that a UMTS user could visit Victoria Peak, call his girlfriend and tell about the trip and show the beautiful scenery as it opens up from Victoria Peak to Hong Kong's skyscrapers and harbour. Nowadays that can only be described verbally. In the future people will use the video camera functionality of the UMTS terminal to increase the experience for both caller and called party.

The shopping experience will also be easier with UMTS. When returning back from the above trip how much easier it would be in say the duty free perfumery at the airport to call your girlfriend and show to her what the special offers are and ask which one would she like. This is even more important when purchasing clothes for your partner. Women have this marvellous knack of being able to bring back clothes that not only look good on their men but actually fit them. Men on the other hand need some remote advice to ensure the purchase does not result in disappointment or even an argument on their return. This service has numerous uses once users are familiar with it. For example when fishing, a man gets the biggest trout ever and calls to his spouse excited to tell about it. Naturally the spouse suspects that her husband is overstating like usual, so he sends pictures of the fish to the spouse to prove his case.

Show Me business uses (remote eye)

The same show-me idea will work in numerous business uses, for example the PC (Personal Computer) repairman might be able to ask the caller to switch on the camera and show what the PC is doing that is causing the problem. The PC repairman might immediately see what the caller would never think of explaining, and in this way deliver very much faster and better customer service by using a 'remote eye' to the site.

The mass market

Video telephony has not taken off in the mass market to date and the main reasons for this have been the immature technology, lack of critical mass of terminals and lack of convenience. When a critical mass of video enabled UMTS terminals is reached, this will change and user behaviour patterns will also change. A key driver for the take up of video telephony will be the enthusiasm related to new innovations and the social pressure that follows the adoption by the early users. Typically this will happen when the penetration of UMTS terminals exceeds 30% of the population or segment. This has been the experience from 2nd Generation systems like GSM, for example in the growth of SMS services.

Video telephony is likely to be more segment orientated in the early years. It will be more used by the corporate segment than by the over 50's segment for instance. It could even be that the youth market could be an early adopter since they take to new ideas more easily and the peer pressure is greatest in this segment. Today in countries like Finland the penetration of mobiles with 15–19 year olds is over 90% and who wants to be the only kid in class without a mobile phone? Likewise who want to be the only kid or executive in the office who can not make or accept a video call?

One of the main drivers for using a videophone will be personal calls when you have the need to have a more personal or 'closer' experience than that offered by PoV (Plain old Voice). It is assumed that penetration will increase steadily, maybe with services like click-to-talk to a support person when help is needed as one of the early take up areas. Usage should grow steadily based on these kind of services being highest initially in the business segment, mainly with video conferencing and then growing penetration and usage by the consumer market as stated maybe with the youth segment possibly leading the take up.

Target segment(s) for video telephony

Video telephony is perhaps the most impressive or 'sexy' of the actual UMTS services and a lot of debate has taken place on whether Video Calls will become mainstream services or if they will end up a marginal service or even just a forgotten curiosity. Technically, video telephony will be possible early on in UMTS but it is unlikely to become very popular first off, as a single video call connection consumes a lot of the available bandwidth on the network, and thus is likely to be rather expensive. Many of the early handsets may not fully support video telephony. One view is that early on it will be a service that will be targeted at business users on the move. Business users will be prepared to accept the higher cost for this service when the occasion arises and even if it is a relatively costly service when compared to a phone call, a video call is much cheaper than air travel to meet face-to-face.

Alternatively some operators could consider that since at the launch of UMTS there will be plenty of bandwidth, video telephony could even be priced competitively to selected non business segments so utilising the early excess capacity. It is possible that this could 'kick start' the video telephony take-up of a service that some still view as too expensive in the early phase of UMTS for the mass market. It could be possible to make video telephony appeal to the mass market by offering the service with reduced bandwidth and cost, albeit with reduced quality compared to what would be required by professional/business use. This theory would need to be tested before it could become part of the operator's overall service strategy.

6.5 SIP (Session Initiation Protocol)

SIP (Session Initiation Protocol), is a new signalling protocol for UMTS which allows a vast range of new services to be created. SIP was set by an IETF (Internet Engineering Task Force) and has been selected by cable and UMTS industries to be the sole call signalling protocol in cable and the 3GPP standard of UMTS. OSA (Open Service Architecture) will also become an important feature.

OSA (Open Service Architecture)

OSA defines API (Application Programming Interface) to provide an open interface to Service Capability Features. The intention of OSA is to open the networks so that 3rd party software applications can be used to build services for the subscribers. Also, service vendors can build applications that can use the service capabilities of any network supporting OSA. Most UMTS services, if not all, could be implemented using this OSA approach.

Making services using SIP

Using SIP as the signalling protocol and OSA as the service creation and execution environment of UMTS networks provides a tremendous opportunity for the operators to really differentiate themselves as service providers. New and innovative services can be created rapidly and customised by each end-user to their personal needs and preferences thus generating customer satisfaction and loyalty. Below are some examples of the possible services.

Richer voice call control

New services start from the area of call control. In practical terms these services include such features as delivering ringing tones and caller image when calling someone. So apart from the current way of selecting how *my* phone rings, with SIP I can now affect **how *your* phone rings** when I call you. Better yet, I can send a picture to your phone when it rings. Thus the subject of the call can be displayed immediately to the caller, for example this call is about Joe's birthday party. Internet URLs (Uniform Resource Locator – the http://www.myname.com addresses which internet web pages use) can be passed within signalling. E-mail and media-on-demand integration can be made more seamless within the call. During a call a video file may be requested, the spoken language can be specified when calling a help desk. And Java or HTML (Hypertext Mark-up Language) payloads may be sent within SIP to support Java based gaming for example. These create numerous combination abilities to use and call up various applications and content and combine those with the call or data transmission. These in turn will allow wholly new business ideas and business models.

9 Vignettes from a 3G Future

Is there Milk in the Fridge?

I am single for the weekend and am in the grocery store. I completely forgot to look in the fridge to see what I need. Luckily I have the new type of fridge which has a cheap simple IP camera inside with a simple remote that can be used via PC or UMTS phone. Now that I am at the store, and have forgotten what I have in the fridge, I can just connect to the fridge camera and take a look.

There will be many devices which can enable and connect to UMTS services. Typically the home fridge, when it will have a built-in camera, will probably be connected via the fixed phone connection. Probably the fixed line operator or high speed data operator, or cable TV company will provide very inexpensive fixed access to such devices. But the compelling part creating the utility for the service is the convenience of access to the fridge camera from anywhere with the UMTS phone.

Richer voice service examples

The examples are far too many to list comprehensively, but some illustrative examples are worth mentioning. SIP allows me to attach the Star Wars theme as a ring tone for my outgoing call attempts, and with it I could add the edited image of me as a Jedi Knight holding a light sabre. If the caller is my boss and the time is after 16.00, forward the call to my voice mail. If the size of in-band image is bigger than 30 KB and it is from unknown sender, do not deliver it to my UMTS phone, but rather re-route it via e-mail it to: me@hotmail.com. If the caller is Stefan, send soccerresults.html file to him without having my phone ring at all.

IN (Intelligent Networks) like basic services

In the 1990s fixed and mobile telecoms operators invested into sophisticated centralised service creation platforms call IN systems (Intelligent Network) to deliver advanced services. Now most of those services can be replicated on standard UMTS services if the operator so desires. These include Caller Identification (Caller ID) Delivery which is also known as CLI (Calling Line Identifier), Caller Name Delivery, call forwarding on busy, call redirect etc. In UMTS these services can also be controlled at the terminal or via proxy if the service is constructed in a way to allow this convenience.

Other SIP uses

Many other services can be created such as **event notification**. It could take the form of an **Automatic Call Back Service**, in cases of high cost incoming calls – when travelling abroad for example, or if needed for network access security systems. It can also be a pure notification service that something has happened. If my son turns on his phone, let me know that he is awake.

Personal profiles may be loaded into the network and used in **dating applications**. "My name is James. I don't smoke. Tall in height. Dark hair. Blue eyes. My likes are: Eating out, good conversation, watching spy movies, technology enabled gadgets, fast cars, beautiful women, dry martinis shaken not stirred. available for Chat wherever I am in the world." Combined with event notification it can take the form of: when matching profile found, send notification and start a chat session. Multi-user games can benefit for example in the invitation: "Sharon is inviting you for a Tetris game session. Join?"

Presence

Services can also use the presence aspect of SIP. Presence includes such conditions as idle, in a meeting, travelling, in/out of the office, beyond network coverage, etc. These can be queried before making a call. So for example if I call Joe and he does not answer, I could send a query and find out that he is in a meeting, then I could send a text message to convey the urgent message. But if the presence told me that he is beyond network coverage, I could contact his boss to try to get a quick answer to the matter.

ASI (Access and share information)

ASI (Access and Share Information) service provides the capability to access information during a phone conversation and share that information between participants in the call.

For instance, a salesperson could browse a product catalogue and price list while having a discussion with a customer. Further, the salesperson could send information like figures, feature lists, prices, etc. on products to the customer during the call.

Click to talk (call-to link)

A Click to Talk service allows WWW (or WAP) pages to contain 'callto' links similar to the way most current internet pages contain 'mailto' links nowadays. When a user clicks a 'callto' link, the terminal initiates a call towards the SIP, URL or number provided in the link. It is possible and beneficial to include additional information to the INVITE message sent by the UMTS terminal when establishing the call. If, for instance, the URL of the page, from which the call was initiated, is included in the INVITE message, the receiver of the call knows immediately the context of the call. In other words, if there are hundreds of pages with click-to-talk links pointing to a mobile merchant's sales hotline, it is very useful for the hotline Call Centre representatives to see exactly which of the many pages triggered this particular call. In the Call Centre the customer service person knows what product is of interest to the customer as well as the customer information and this can be displayed on the screen as the call is delivered.

SWIM (See What I Mean)

SWIM (See What I Mean) service provides the capability to illustrate verbal message with drawings or figures. This could be a shared white-board during a phone conversation. During a call a map of the current location is retrieved from a server and both parties can draw marks on top of that, thus giving instructions how to find to the agreed place.

IP Multimedia calls will provide for a myriad of new services that will obviously make lives more interesting and easier for the users. Some of the above examples bring about a realisation that there will be many occasions during normal lives both work time and personal time when people will use new UMTS services when they become available.

6.6 Social messaging

Communication is an essential part of our lives. People find all kinds of ways to communicate, not least with family and friends. Telephones have been a standard form of communication for many years and now with the advent of the Internet era, people have been surprised to see the Internet not being solely used for data information. The Internet brought the phenomenon of 'chat', with people communicating in real time with PCs. Now with the mobile era, people no longer use their mobile phone for just verbal communication. Mobile communication has now heralded the age of social messaging between people only using mobile phones. The earliest form of social messaging was SMS text messaging, and soon after its advent, mobile chat emerged.

Once again, it has been the young that have adopted this new technology. In many countries, mobile calls have been traditionally more expensive than fixed ones. For many young people, there has been the need to balance limited sums of money with the latest 'fashions' in technology and clothing. When the cost of mobile handsets dropped with mass-market use, it was only a matter of time before the young became their 'champions'. To expand this market segment, many mobile phone companies adopted market techniques to increase sales with the young (e.g. colourful, individual and changeable covers for mobile phones). However it was not for voice calls that the young have increasingly used their mobile phones. SMS (Short Message System) has been the revolutionary use of mobile phones with the young.

SMS (Short Message System) or text messaging

SMS is the text messaging system that can be used between mobile phones. It is simple cheap and effective. The messages take seconds to write and can be normally sent to any other GSM phone user. In December 1999, The GSM Association reported that the worldwide SMS traffic was a staggering 3 billion messages a month. This rise had been dramatic considering it was 1 billion per month in April 1999 to then 2 billion in October 1999. In August 2000 SMS traffic had risen to 9 billion messages per month and the growth appears to be limitless.

Mobile chat

SMS relies on the individual messaging a recipient and waiting for the other to respond. It is more limited when comparing to an Internet chat room. SMS normally occurs between individuals, not groups. In comparison, Mobile chat is very different in its nature. For those not familiar with the internet specific meaning of the term 'chat' we do not mean talking or conversation, but rather written discussion. Internet chat is concurrent, real time live text-based discussion via a messaging board, where all who participate type in their comments, usually one or two lines in length. All comments are displayed in the order they are entered, producing dozens of consecutive comments, sometimes hundreds of lines of text. Often numerous 'threads' of discussion take place within one chat session where a pair of people might talk about the actual topic, and another pair might be flirting with each other, etc. Also typically there are some who only observe and read messages without writing, while others are very active participants.

With mobile chat the mobile user can communicate with dozens of people at once. They can also set up 'friends lists' for more frequent recipients and have several chat conversations at once. This is more in the style of the more commonly known Internet chat rooms. Internet chat rooms allow a person to enter real time conversations wherever they reside in the world. They can choose to chat about any subject they choose and can normally specify what geographic areas they wish to make contact with (to keep the language and life experiences to a similar level).

Another major factor in the growth of Internet chat rooms is anonymity. Anonymity meaning that the person can choose to enter a conversation and not reveal their true name, location and personality. People can therefore talk more freely and then choose with whom they wish to have further contact. Other factors that have resulted in the growth of chat has been the low cost of the service and the friendships that can be established.

Internet chat rooms are normally free to enter. As in any fixed internet services, the caller may have to pay for the call charge and Internet access in their respective country but especially in the US where chat is booming local call charges are free. Once 'chatting' in the various chat rooms, it is then easy to make contact with people of similar interest and this is easy as the person chooses what type of room they enter. For these reasons it is not surprising that it is the young that were the early adopters of the Internet chat room. As mobile chat offers all of the same features plus mobility, we can predict that the young will pioneer the use of new mobile chat services.

Today with a mobile phone, anyone can enter this new, interactive, mobile group communication. Rather than the limited chatting with SMS or the 'fixed' aspect of email, mobile chat allows active and simultaneous communication. It allows this to happen wherever the person happens to be and only whenever or wherever they decide to be on line. People can still remain anonymous and mobile chatting continues to be much cheaper than maintaining a phone conversation between groups.

Community nature of mobile chat

Perhaps most importantly, mobile chat allows the creation of new 'communities' or groups of people with similar interests. People can quickly establish lists of 'friends' who have similar interests like cooking, football, sport, music, culture or movies. When something significant happens within their sphere of interest, the group can quickly and easily make contact. This could be arranging the team for next Saturday's football match or an outing to the movies or general chat as to why the soufflé went all wrong.

When needing to arrange something, it is quick, easy and cheap to be in contact with others whom the caller has links to in either private, family, friends or public chat rooms. Despite the sudden and explosive growth of mobile chat, textual messaging is only the start of these mobile services. With the growth in technology and UMTS, text chat will give way to pictures, sounds and video trailers all being added to the mobile conversation. Not only is mobile chat here to stay, it will probably revolutionise the way we use and consider our mobile phone.

6.7 Standards and protocols to help customise services

UMTS is about services. The Me attribute is the most personal and the one valued most by the end-user. End-users will want their services ever more personalised and customised. The UMTS environment provides several protocols and standards to create further customisation of services and while this is not a technical book about UMTS, some of the major standards and protocols need to be discussed to illustrate some of the remarkable flexibility which UMTS will have to personalise services.

IP telephony

IP telephony (or IP multimedia) enables flexible service creation for multimedia services allowing operators greater flexibility to create their own integrated applications, which dynamically combine a variety of media types. This will be one way that operators will be able to differentiate themselves from their competition.

Integrated IP services

Some examples of these integrated services are video telephony between IP terminals, application sharing during a voice or video call (showing pictures or documents currently discussed or image of a product or place), m-commerce (pushing product picture and information to the customer during call) and any type of click-to-dial services. Although many of these services are already possible with circuit switched services and WAP/WTA, IP as a platform offers more flexibility in service creation and execution and UMTS offers the capacity needed for smooth introduction and management of the operation.

In traditional mobile telecommunication systems, most services are standardised in detail and limit the operator's ability to create differentiation in their service offering. Operators would however prefer a more 'tool-box' type of service environment, enabling unique service creation. With increased global competition in UMTS, not only from global UMTS network operators, but also from MVNOs (Mobile Virtual Network Operators), service differentiation will be one key to success. The competitive dimensions of UMTS will be discussed later in the Competitiveness chapter, but IP services are one way to differentiate from the simpler WTA and WAP standard services. Better yet, the unique IP services should be available on a global scale instantly. Furthermore, they should behave and appear to the user in the same way across terminals and networks.

My services with Me (or Virtual Home Environment – VHE)

The VHE (Virtual Home Environment) is defined so that users are consistently presented with the same user interface customisation and services in whatever network and whatever terminal (within the capabilities of the terminal and the network), wherever the user may be located. This allows the network provider to create services which travel with the user onto other networks through 'roaming' and still retain the look and feel of how the services appeared and behaved at home. A good example is the Japanese character set in the user interface. Certainly a UMTS operator in Norway could try to emulate the Japanese command structure of services and whenever a Japanese businessman or tourist travelled to Norway, offer the Norwegian services translated into Japanese and displayed with Japanese characters, etc. It is quite likely, however, that the Japanese traveller would prefer to see his familiar service on its familiar character set and seek out Norway-related information rather through it. This is what VHE was designed to accomplish. Of course the reverse would then be true of the Norwegian in Japan.

6.8 Profile management

As our terminal and UMTS service become ever more personal and important to us, we will want the personalisation to be replicated across multiple devices and services, and we will want to have control of our own profiles. We may want numerous profiles, such as one for work, one for our family, and one for the fans of the football club we support. A User Profile Server or Application is way to accomplish this by enabling each subscriber to define a set of profiles covering all of their network, application and information services, with easy switchover between profiles as needed. In order to manage their business and private lives business end-users will require the ability to define a set of profiles covering all of their network, application and information services, with easy switchover between a number of profiles. Profiles for example will be different depending on time of day or day of the week as people move between work and leisure time. Typically, a user would create a profile for each role in which they operate, including roles at work and personal roles; and for different geographic or lifestyle situations in which they operate.

Typically, a subscriber would create a profile for:

- each role in which they operate, including roles at work and personal roles
- different geographic, language and lifestyle situations in which they operate
- information needs
- company internal communities with which need to interact
- deputy structures to allow automated re-routing of urgent information to default deputies in case of illness, vacation, trips etc.

As part of each Service Package, the Profile Application should provide self care and profile switching for all services provided within that proposition. The services to be customised by a Profile Application should support an asynchronous self care interface as well as the normal interactive interfaces.

6.9 *Me, myself and I*

This chapter has examined the way services can address the Me attribute of the 5 M's (Movement Moment Me Money Machines) of UMTS services. It is important to allow services with the Me attribute to extend the self into the community, to build ever more attractive services for each individual in the UMTS network. In building friendships, relationships and communities the UMTS user will want to make others feel good and happy.

Me is the most personal of the 5 M's and it is valued most highly by the end-user. It can be used for good to make services ever better, and it can be used for bad if the trusted personal device offends us or breaks our trust. Operators, application developers, systems integrators and content providers need to pay particular attention to the Me attribute to ensure that services are perceived as good for the individual, and allow for rapid corrections if a given service 'improvement' or change ends up being badly received by the users. With all this there are bound to be many lessons learned through trial and error, but a positive attitude and a touch of humour will help smooth out the rough spots. Perhaps a guiding thought could come from Lina Wertmuller who said: "Laughter is the Vaseline that makes the ideas penetrate better."

7

'The best way to make money is to make money.'
Old saying among money forgers

Services to Address Money Needs:

Expending Financial Resources

Tomi T Ahonen and *Joe Barrett*

The ability to spend money or use the mobile terminal for money transactions is another useful aspect of services for UMTS (Universal Mobile Telecommunications System). Making purchases and having those billed onto the mobile phone bill is one unique aspect that UMTS operators can build upon and create differentiation. The UMTS operator will be the only outlet that will be able to bill for very small payments electronically and this will more than likely become very desirable for a large percentage of society. Additionally, mBanking and other financial transactions via the UMTS terminal will develop into useful everyday services. Operators should consider what are the recognition factors, the secure and tangible benefits that will enhance the status of the mobile terminal as a reliable and trusted device.

The services described in this chapter have a high benefit on the Money attribute. There are hundreds of such services and we have a chance only to explore a few of them. But to illustrate by way of example some of the obvious uses of Money type services, they include:

Accessing cash machines	Online services
Advertising	Online stores
Auctions	Paying bills
Betting at sports event	Paying government fees and taxes
Betting in general	Promotions
B2C (Business to Consumer)	Sending money/lending money
Coupons	Sponsorship of content
Currency rates	Stock market information
Free trials and demos	Stock market trades
Gambling	Ticket purchases
Locating cash machines	Trading and bartering
Lottery	Vending machine payments
mBanking	Yellow pages

This chapter will discuss some services which have a strong benefit from the Money aspect of a UMTS service. The services are not in any order of importance and this brief discussion will not be able to adequately address even the major areas of Money type services. But a deeper discussion of a few services is useful to understand the nature of the Money aspect of UMTS services. As mentioned in the introduction chapter, all UMTS services should be designed with a Money dimension if at all possible.

7.1 Mobile commerce (mCommerce)

Mobile commerce is one of the biggest opportunities for the mobile operator. Existing at the very heart of UMTS, mCommerce is that type of mobile commerce which very easily addresses all five of the 5 M's (Movement Moment Me Money Machines). A service which is adjusts for movement and time, becomes personalised, is automated with machines, and of course deals with money. For example commuters purchasing train and bus tickets would be close to an ideal such service. The service involves something which is not always tangible and could be embedded into another service or enable the extension of a service such as purchasing a travel ticket. The need moves about – we board the train from a different station going to work, and going home. The service can become time-sensitive, such as recognising patterns of travel and automatically offering a discount ticket, or monthly pass for example, when the travel justifies it. As part of a promotion service the mCommerce or Money characteristic of the service allows the main service to add more value to the end user. In this case it could be the redemption of a discount coupon for rail travel.

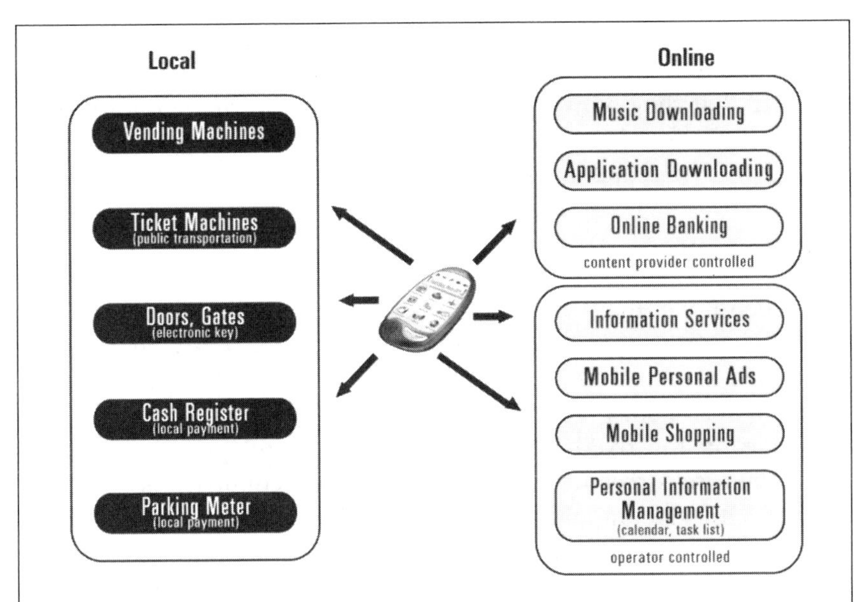

There will be two basic access types for mCommerce, local and online. Local connection can be either via Bluetooth to machines like parking meters or cash registers to pay for goods at point of sale or online to sites where goods or services are paid for and then downloaded to the UMTS terminal. This could be music or ring tones. Alternatively the goods are paid for and then collected or delivered by the postal or courier service.

B2C (Business to Consumer)

At the core of B2C (Business-to-Consumer) services are sales and services to consumers. To a large degree the services discussed in this chapter can be all categorised as B2C services. Mobile commerce can be concrete items we can purchase at the store such as paying for soft drinks bought from vending machines, or even making regular store purchases and using the UMTS phone as a mobile wallet. The mobile wallet idea can be seen more along the lines as a credit card or bank/debit card.

7.2 Buying and consuming digital content

An ideal mCommerce service is anything which can be consumed on the mobile terminal. This might sound like a hopeless idea for some, thinking that they would not *want* to consume *anything* on a mobile phone. But a few examples will help illustrate the opportunity. The other 5 M's help find a lot of opportunities.

Buying a map

Movement: we may want to purchase a map to find our way around a strange town or country. Maps currently are mostly consumed as paper printed maps, or sometimes as books or booklets of maps. Today with the digitalisation of maps they are freely available on the fixed Internet so access to and printing of maps even down to street level in most developed countries is relatively simple. Others companies provide CD Rom based mapping software and this is used already in many car navigation systems. The point is that currently we consume a map in visual, often paper-based forms, and often pay for the map. There is no reason why the map cannot be sold to us onto our UMTS phone. The beauty of the purchasing of a map on the UMTS phone, is that we can act upon the impulse the moment we are 'lost' – we don't have to go looking for a bookstore or tourist shop. The map we bought is not back at the hotel or forgotten at home, or in the car's glove compartment. As we carry our mobile phone with us at all times, the map – or the ability to purchase a map – is with us at all times and in all places. Mobile maps are one of the distinguishing services on the J-Phone service in Japan.

Crossword puzzle

Another example is that when we find ourselves in a sudden or unexpected delay we often have time on our hands. For example if the plane or train is delayed. At such times we might go sit at a café and buy a newspaper, do the crossword puzzle or read the sports scores. Again, while we are not accustomed to consuming sports news or doing the crossword puzzle on our mobile phone, there is nothing preventing us from doing that. And we are accustomed to using some of our spare time for such leisurely and relaxing activities through the newspaper. However, the UMTS phone provides us with numerous benefits over the random time-killing need. Our UMTS phone is always able to deliver our content, we are not dependent on finding a news stand, and its hours of operation. In foreign countries we might not have access to our favourite newspapers, but our favourite content can always travel with us on our UMTS phone. And the service can be more up-to-date than a printed newspaper which we all know is delivering 'yesterday's news'.

Car racing game update

The Me attribute, the ability to provide personalisation according to our needs and the expansion of the personalisation to our communities brings about many opportunities for mobile commerce services. Our preferences and profiles can help find many traditional brick-and-mortar store sales opportunities, such as targeting a tennis goods special at a tennis playing person or golf gear specials only to the golfer(s). Wherever there is existing consumption of goods and services in digital format, which can be readily used in small portable devices, then the ability to target customers benefits directly from the UMTS environment. The typical optimal applications are games and music. If the game developer has the UMTS phone customer already in its user database for a racing car game, then when a new game is developed – perhaps racing boats, or racing airplanes – they can directly target those customers who have already purchased similar game software before. Click once to purchase, enter your PIN (Personal Identity Number) code and it is sold and downloaded in seconds.

Lending money

The Money attribute suggests a lot of options and the easiest to identify is the abstract notion of funds. Money, expressed currently in various formats from cash paper money, to cheques, to balances on bank accounts, to credit lines on credit cards, etc., is a natural to be consumed and transferred via UMTS. For example lending your friend some money. With UMTS lending your friend 20 dollars will be as easy as taking the notes out of your pocket and will be just as simple as sending a short message to your friend's mobile phone. This is in fact already happening in some markets using existing technology and we will see the continued integration of money and mCommerce services so that physical money is replaced by electronic 'bits'.

Buying software

The machines attribute brings yet another type of digital consumption of mobile commerce on the mobile terminal. The best example here is the direct download of new software and applications. So if you would like to purchase an update of some software, for example a fancier version of image manipulation software, that can be directly purchased from the manufacturer's automated website, and downloaded and configured directly to the device. The software manufacturer saves in countless steps from marketing to storing onto CD Rom to packaging to shipping to fraud prevention; while the user receives immediate download and use of software.

The key to all of the above examples is that there is a mobile commerce opportunity in areas which are already using digital content, especially if that content can be readily consumed on the mobile phone.

Digital rights management

It is important at this point to bring up digital rights management. The software companies, the music companies, ring tone suppliers, and all companies that supply a product in a digital form to the mobile terminal will require assurances that any copyright will not be violated and royalty payments lost. There have been moves to slow

10 Vignettes from a 3G Future

Birthday Cam

The birthday boy is surrounded by his friends and approaches the cake. 10 candles, the big event. If only daddy could have been here. Luckily mother can set up the UMTS phone and call daddy who knew to set this time aside on his business trip, so that he can also join in to see if all candles are blown out.

The spontaneous contacting needs and now the ability to see the important people and their important events as they happen, are going to provide countless special moments where the UMTS phone connects people like no other device before. The UMTS operator will need to enable various easy ways to experiment with these new communication means and price them attractively so that usage will become commonplace, and the change in behaviour can start to happen.

down the availability and distribution of digital content, mostly because it was being distributed free on the fixed Internet. But once the relevant industries are confident that a secure business model exists in UMTS that enables them to make money, they will move quickly to enter the market.

Security of the mCommerce service or the security of the copyright of the content is a major requirement and will need to be in place before there is mass market availability of certain content. So as the Money attribute is considered as part of the service offering it is not just that it meets a money generation or financial kind of service, it may also have to meet specific security and digital rights criteria. These have to be part of the network and terminal capability.

7.3 Intangible services

While the ideal mCommerce purchase involves something which can be consumed on the mobile phone, there are also purchases of certain services which are genuinely intangible. The obvious examples are mobile banking (discussed in the Movement chapter), insurance and ticketing.

Ticketing

Ticketing is likely to be one of the early success stories of mobile services. Several other examples mentioned elsewhere in this book talk about ticketing in particular ways, but ticketing should be viewed overall, considering that any tickets sold anywhere are great prospects for UMTS services. The service being bought is of course not the ticket itself, even though in some cases the ticket itself may even hold residual value as a collectable item say for a farewell rock concert. We buy the ticket to prove our right to do something, such as fly on an airplane, see a movie, take the bus, enjoy a ride at an amusement park or attend the theatre. The event, occasion or trip, is what we are actually purchasing. The ticket is only our proof that we have paid.

For those enthusiasts who actually want to have a physical ticket as a memento, a mobile ticketing service is in their eyes a backward step and one to be resisted. For the event organiser however it is an opportunity to sell souvenir paper tickets for a small fee. Alternatively the collector may be more inclined to purchase a program booklet as the memento.

Seat selection

The first benefit from UMTS based ticketing is access to information and tickets without standing in line. Much like fixed Internet based ticket sales today, the UMTS terminal allows us to view seats, prices, even possibly what the view might be like from that part of the stadium. This is one way to sell those 'best' seats at a higher price. If the customer can see the actual view they will have of the stage or the screen, they are more likely to take the seats with the best view at the higher price. Current, paper based ticketing solutions tend to force tickets into strict groups where all viewers in a given stadium section pay the same price for their seats, with the next expensive tier of seats costing typically 10–20 dollars more. The customisation ability of UMTS billing and service creation allows each row of seats to be priced separately, which could have increments in seat billing of a dollar or even tens of cents per row. By monitoring which seats sell out fast and which slowly, the event organiser and venue owner can build an optimised *dynamic* seat pricing plan which learns actual preferences and adjusts to maximise revenue.

All UMTS based ticketing brings significant improvements over other electronic ticketing available today, and specifically those of the fixed Internet. This starts with payment. Here we do not have to send separate credit card information to the ticketing company. The event or trip or amusement ticket price is added onto our mobile commerce purchases on our mobile phone bill, neatly itemised and dated.

Avoiding standing in line

But the most compelling benefit of UMTS ticketing is the lack of any need to stand in line to pick up tickets. This aversion to standing in line is highly evident in the UK where there is even a company that provides a queuing service, for a nice fee, so that long queues can be avoided. With UMTS ticketing, the ticket authorization can be linked directly to the mobile phone, and its validation can happen electronically. In this way there is no need to arrive to pick up tickets, nor to show the tickets. Being totally electronic, the process of selling, billing and validating the right to the ticketable event or occasion will be vastly simplified as will be the costs of managing the money collection.

7.4 Brick and mortar store purchases

There will be plenty of opportunity to use the UMTS environment for selling products and services already available at bricks-and-mortar 'real' stores on the high street. This could be anything from buying books to clothes to second hand cars and homes. There are of course many limitations to looking for and deciding on a purchase on the smaller screens and keypads of UMTS phones, where it is of course rather unlikely that people will actually make a decision to buy a house only via the mobile phone.

Finding a home

However, *renting* a flat via a mobile phone can become quite viable and would save a great deal of time. *Searching* for a flat via a UMTS phone might be much more common than via a newspaper only a few years from now. The 'for rent' or 'to let' advertisement on the UMTS phone can gain from several particular benefits that newspapers cannot deliver, such as location – show me the apartments for rent within walking distance from this underground train station, or personalisation, show me only those apartments which are not on the ground floor or in the basement.

Of course the benefits mobile commerce bring to purchases of real goods from real retailers can include those that are associated with eCommerce and mobility, such as price comparisons, catalogue browsing, variety, special goods, and so forth. As the goods will need to be separately delivered, there is a diminishing return on what it is really useful to order via a mobile phone, although there have been experiments yielding positive results on buying anything from personal computers to ordering pizza delivered to your home.

Try-it guarantees

One of the most difficult questions to wrestle with is will customers purchase goods via their mobile device? This would mean possibly looking at and checking out say a new pair of trousers or dress or new dishes for the kitchen. The catalogue business has been working successfully for many years with a model that includes the option of delivery, try and buy (return if it does not fit) without having to leave the sofa. So the business model is well established. The question only arises from the credit viability of the customer. In the catalogue company business model, before the catalogue company will extend credit there is normally some form of credit check. This covers the risks of people trying and not paying. With post-paid subscription mobile phone users this may not be an issue since there will be a transaction history that the operator can refer to. A credit rating can be extended to the retailer offering the company a greater level of confidence in the purchase. However pre-paid customers pose a risk since operators have less information on them, especially the home address which is necessary for a home delivery, try then buy service.

This may not be a bad thing. Since operators are looking to move existing pre-paid customers to post-paid billing, offering a catalogue kind of service only to post-paid customers could encourage this migration. And nothing prevents a pre-paid customer from providing credit card information as back-up, or filling out an on-line credit form for 'big ticket' purchases.

Whatever the strategic choice in delivering mCommerce services a UMTS operator selects, the chosen strategy should leverage existing assets, be they tangible or intangible, and should probably build on the proprietary knowledge of the customers that the operator has. They should also leverage the unique characteristics of UMTS.

Vending machines

A very specialised niche market is the vending machine. For most travelling people the notion of having appropriate coins in different locations or countries is a continuous problem when there is a need to use a vending machine. Vending machine operating merchants have of course the opposite problem of continuously sending people to unload the cash that has accumulated in the machines. For both parties it is more useful to only authorise the payment via a wireless connection using the mobile phone, deducting the payment from the customer's mobile phone account, and adding the money to the account of the vending machine operating merchant. It removes the problem of having to have the appropriate type of coins or banknotes from the user, and reduces the frequency of having to send people to empty the cash from the vending machine by the merchant.

Here the operator can act as a broker to enable the transaction and benefit both parties, taking a small percentage of the sale for that work. Naturally the consumer will want to see an itemised bill of what has been consumed on the mobile phone account, so the individual purchases of soft drinks, candy, tobacco, etc., from vending machines will need to be itemised on the mobile phone bill.

There is also the requirement to ensure that any transaction is properly authorised. It is not ideal if your child plays with your mobile phone while you are at home, hits repeatedly the last number redial on the phone, and suddenly somewhere in the city a vending machine starts to dispense cans of cola drinks for free to passers-by.

7.5 mAd (Mobile Advertising)

Perhaps the most heated topic of new services, and definitely the one arousing the most passionate opinions, is that of mobile advertising (mAd). The often expressed view is that people do not want advertising on their mobile device. In part this is the industry's fault since there has been story after story about how UMTS users will walk past shops and get the latest promotion or sales blurb appear on their phone along the lines of "we have Levis jeans on sale today". "Wonderful" say the operators and analysts – "no thank you" reply the consumers!

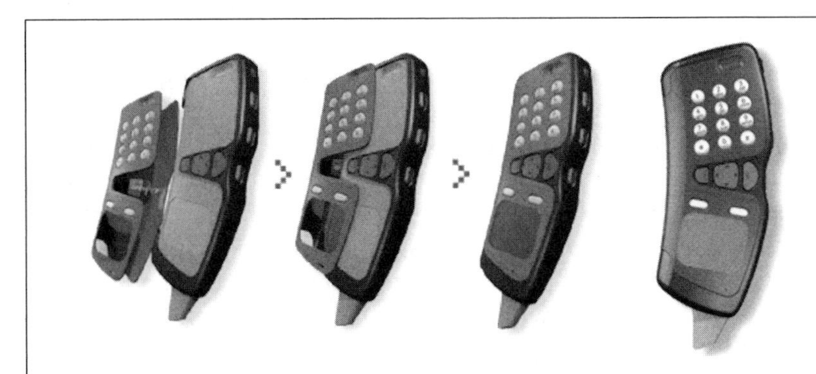

There are many variations of new phone designs. Gitwit have one interesting view with slide on covers and the screen and antenna at the bottom of the phone.

Fear of mobile junk mail

The experience of junk 'snail-mail' – or what used to be called junk mail – carried to us by the postman both at home and at work, has taught us to be wary of this form of advertising. The advent of junk e-mail and now the prospect of junk mobile-messages is enough to keep users at home, under the covers of their beds with their mobile phones turned off, where the mAdvertisers can not reach them. It is important to recognise the perceptions that users have about advertisements. On TV it is easy to flip channels when the adverts interrupt the programming. In newspapers or magazines adverts are often overlooked or they are so visually stimulating that the readers take time to scan the messages. Bill boards and posters permeate our external environment. So we are actually surrounded by advertisements all the time. In reality they are an everyday occurrence in our lives. Advertisements are often mass-marketing, and thus we are bombarded by thousands of distinct marketing messages through dozens of channels, and if we are lucky, one will actually seem meaningful to us on any given day. The rest are seen as an annoyance.

The mAd approach will need to be carefully designed to encourage user acceptance through sensitive delivery of promotions and advertisements. These should be minimum at the start of the service, growing in number and relevance as the operator builds up a more comprehensive user profile.

So many early comments on the advent of advertising on mobile phones have been those of outright rejection, of concern, and of doubt. As the sentiment seems to suggest that individuals do not feel comfortable with the idea that mobile advertisements will start to flood their inboxes on the mobile phones this area of the business needs to be looked at from a new perspective. If there is fear that some of the worst examples of junk mail at the home mail box and spam e-mail at the fixed Internet will arrive on the mobile terminals it has to be negated or ideally slipped in by the back door while no one is looking.

Whatever the approach and strategy taken it is vital to understand the customer concerns. There have been early stands on the ethics of advertising on the mobile terminal, such as the ethical guidelines issues by the WAA (Wireless Advertising Association). A few significant observations need to be made overall.

mAd will find its place

First of all we need to keep in mind that all current media are supported by advertising in some form or another, and when considering promotion, many non-traditional forms of human communication are also supported by money from third parties. Advertising is a significant revenue stream for television, radio, newspapers and magazines for example, often heavily subsidising the cost of the content delivered. In a more limited form, advertising exists in books – the authors mention their other books at the beginning of the book, and publishers sometimes blatantly advertise other books at the back covers or back pages of books. This is especially seen in paperbacks.

The Hollywood motion picture industry has both in-movie **placement advertising** (such as James Bond's brands of cars, champagnes, watches and mobile phones), as well as a lot of sponsorship seen sometimes only through the credits at the end of the movie. Lets not forget the 5 or sometimes 10 min of promotions at the beginning of rental video tapes advertising the upcoming releases. These in no way subsidise the purchase price of the VHS tape for instance. So far there has not been a similar approach taken with DVDs since unlike VHS tape, DVD is non linear. We do though expect that these 'upcoming promotions' will soon appear as part of the menu alongside the 'How the movie was made' and 'Interviews with the stars' so that customers will actually chose to view these specific advertisements.

Event sponsorships

For promotion, we can find advertising at almost all different walks of life. Major sporting events are of course sponsored by numerous worldwide brands, while local sports events are often sponsored by local companies. Sponsorship includes the arts, with promotional grants from major corporations which want to be seen as supporters of cultural events. This includes such areas as sponsoring an exhibition at a museum, or an opera tour. In many cases these events would not exist without this sponsorship. Business events are sponsored by various companies wanting to do business within that industry, such as the telecoms equipment manufacturers sponsoring trade shows and conferences.

It is important to note that advertising and sponsorship is already prevalent and omnipresent in our lives. Advertisers have always strived to find the most cost-effective means to reach their target audiences. They have found that television reaches large audiences but is not very targeted. Direct mail ('junk mail') can be very personalised, but is expensive, and still yields low response rates. Internet banner ads seemed very promising until true statistics emerged on click-through rates and illustrated that the fixed Internet advertising opportunity was not as efficient as initially expected.

mAd will bring about a new opportunity for advertisers. While some will argue that it is the ultimate advertising vehicle, it is quite likely that 10 years from now all other forms of advertising will still remain, and mAd will have become just another channel to consider in any advertising campaign. There will be many lessons to be learned from using mAd and of course there will be mistakes made in learning those lessons as there have been in the fixed Internet.

The mobile terminal does bring several very compelling arguments for why it is an exceptionally powerful tool to deliver advertising. The mobile terminal is personal, we associate ourselves with our phone, we change its covers, ringing tones, and personalise it to have it communicate our personality to our surroundings. In some ways our phone is a virtual representation of ourselves. We don't have any other device which we consider more personal to us than the mobile phone. As it is so personal, we will easily feel happiness from value delivered through our personal trusted device, but also strong anger if unwanted and intrusive things are brought to us via the same device. For all who arrive to this matter from outside the mobile telecoms business, here is a remarkably valuable lesson. Be very certain that nobody in any given target audience can be offended by your advertisement. Consider the impact of advertisements on the television. If we see an offending ad on TV, we rarely complain to the TV station. But if we were to receive an objectionable and unsolicited ad through the mobile phone, we might very well consider changing operators. This situation has to be avoided.

Precisely targeted ads

On the positive side, personalisation brings about the ability to directly target the operator's customer base. Even direct mail (junk mail), e-mail and web banner ads on the fixed Internet have some degree of personalisation. The challenges there are related to bad data being collected or created in user databases where the user is not actively interested – nor rewarded – for ensuring that profiles are accurate. And the ability to target is not precise enough. Internet profiling software that tracks usage from a PC is not tracking personal usage, it is tracking household usage which can consist of parents, children and other family members maybe visiting for a few hours.

With a mobile network every user's phone is separately identified, and of course the mobile operator tracks every UMTS phone's user data separately. This is an enormous database of trivial data, but mobile operators are already well equipped to collect and handle such databases. The current billing systems contain a great deal of data on what subscribers are doing on a week to week, day to day and minute by minute basis. CRM (Customer Relationship Management) systems can contain details of everything from contact details to what was the last content downloaded, from where, what was purchased and the location time and cost of the transaction.

Towards the segment of one

For the advertiser this brings about the ability to truly target, and laser focus their messages to tiny segments, so-called 'micro-segments' and even theoretically approach the ultimate 'segment of one'. Imagine how this would work in the music business. Initially the UMTS subscriber could indicate a liking of popular music, and perhaps subscribe to a music streaming service of a pop music channel originating for example from London. A few months later, that UMTS user might order and download directly to the UMTS phone music from their favourite artist. Now the service provider could offer the customer a special deal which could, for example be based along the lines of how Amazon.com sells books. "We've noticed that many people who bought this music have also recently purchased music from..." and of course offer the chance to listen to a short selections from those artists' songs.

Advertisement of upcoming event to dedicated fans

Now, a little later, when the artist arranges their next world tour, the tour promoter would like very much to target explicitly those who have bought the artists music in the recent past. The tour promoter would like to deliver promotional messages (music information and news) directly to these fans via their mobile phones. "And by the way would you also like to be one of the first to purchase tickets and get one of the best seats in the house?" But even if a given fan does not want to purchase the tickets to the concert, the fan is likely to store the advertisement, and show it to friends and other fans, and even forward it to their friends at their own cost, creating expanded and free delivery for the content provider. This is then not seen as intrusive or offensive junk mail; quite the contrary, fans would even pay for that kind of 'advertising' to be sent to them on a regular basis.

7.6 Adver-tainment?

Now we have a blurring at the edges of advertising services. Where do we draw the line. Are these advertisements or paid-for content? Here is where the traditional concerns about advertising end. If a true soccer fan, who has purchased or downloaded sports content directly to their UMTS phone, gets a personalised message from their soccer idol, the fan will be happy to receive it. They will feel privileged to have received a 'personal' promotional message and will not be offended.

The promotion of events and tickets is not limited to one industry. The same is true of Manchester United football, Ferrari Formula One racing, NY Yankees Baseball, and any area where fans are passionate and pay to see events. These are all very promising target audiences for very successful mobile ad campaigns, and then it can very easily become a reason to *switch* operators to gain access to one's own favourite teams, artists, or events.

Immediate call to action/click to buy

Perhaps the most valuable attribute for the mAdvertiser is, in the final analysis, click to buy. On television, radio, newspapers and magazines, even if you totally convince the target person to want to buy something, they still have to take separate action with another medium to do so. Come to the store, call this number, send in this coupon. With mAd there is the natural ability to make a purchase or an order or contact the company, or take some other desired action – right on the spot of the maximum effect of the ad, exactly at the same time as one is consuming the ad. Click to buy: by placing a click-to-buy link on the ad, the advertiser can achieve with mAd what few other means can provide – immediate sales. This may become the new measure of an ad campaign's success – how strong was its sent ads to click-to-action ratio? The higher the better of course.

Do not kill the opportunity

One final thought to consider about these kind of promotions. The enthusiast, be it sports, music or whatever is willing to pay for a promotion if the related content with it is perceived as valuable, such as news, entertainment or direct personal communication. The promoter is willing to pay for the delivery of the promotion to encourage sales take-up. The operator now has a number of options and will be tempted to take two bites of the cherry and get two revenue steams. The operator must be very careful with this, it is our firm belief that if users are forced to pay anything for an advertisement, they will stop consuming them. That kills the goose which lays golden eggs.

There is a natural reluctance towards advertising in general. UMTS provides an opportunity to make advertising less disturbing, and in some targeted cases even welcome. But the operator should resist taking anything from the subscriber when consuming mobile advertisements, and ensure that adequate payment is received solely from the advertiser. The cost-efficiencies and targeting possibilities in UMTS advertising should guarantee that a significant part of the total advertising spend will migrate to UMTS.

Blurring the line

Of course hybrid solutions can very easily be created with UMTS. A '**what's on tonight**' service could show theatre, movie, concert etc., listings for the town. That could be a billable service, for example as a part of the basic information bundle offered by the UMTS operator. Interlaced with the real information of the service could be paid content sites – "click here to see 30 s clip of the musical" – and there could be totally free sponsored sites – "Free site: click here to see tonight's **last-minute seats** that are still available." The first site would have a legitimate entertainment value – seeing 30 s of the performance – and its data transmission is readily understood to be expensive. The second item is strictly a blatant attempt to sell empty seats for a play which has to run tonight anyway, and any money for a last-minute seat is better than nothing.

Location based push ads and coupons

One of the defining aspects of UMTS services is the Movement attribute, and it sets UMTS services apart from most others. The natural early assumption for this is of course, that we can use mobility in advertising, to get local ads pushed at us when we walk down the street, or at a shopping mall, or driving our cars. While this is likely to become one part of the total mAd opportunity, it is likely to be a marginal part of the mAd equation.

The big reason being that as each ad needs to be paid for, there is really little interest for the local merchant to pay for pushing 'mobile spam' at all passers-by. Imagine a shoe store wanting to advertise its 1-day sale. On the radio there will be a set fee for a set amount of advertising minutes of airtime. On the mobile phone there would be a separate cost for each individual ad. The power of mAd is that it can be focused, and that focusing cancels its individual delivery cost. If the ad is not focused, it loses it power. No matter how much the shoe store is selling its stock at '40% off' just for today, there will be little demand if most of the passers-by happened to have purchased a couple of pairs of shoes last week. But everybody who is *not* interested in shoes is likely to be upset, offended, even outraged by the unsolicited ad, and may hold a grudge against that shoe store, and certainly also the mobile operator for allowing this intrusion. If not by the single shoe store ad, but especially if our mobile phone starts to beep 2 min later after receiving an ad from the nearby toy store, and 3 min after that an ad from the insurance agent's office, then the realtor, the barber, etc.

Targeted ads to loyal customers

But lets imagine it the other way around. The shoe store owner gives a free mobile coupon worth 10 min of talk time at local call rates with every shoe purchase during say the next 6 months. This would in most cases amount to a discount of less than 2 dollars per pair of shoes bought, hardly an overwhelming marketing cost. But as those calls are redeemed, the merchant gains the mobile numbers of its loyal customers.

The merchant might want to try to set up a mailing list and create for example an SMS (Short Message Service) text message based advertisement for those customers. And some innovative and technologically oriented merchants will, as they learn to create in the UMTS environment. In the vast majority of cases, however, the database and advertisement will be handled by the operator – the natural owner of the database, with the ability to collect the data and the solid processes to convert that into an advertisement. In this example, let us assume the store owner wants to run the same '40% off sale' today. Rather than targeting all random passers-by, which could be 3000 people for the sake of argument, the store can now target *its* 3000 customers over the past half a year as the people *most likely* to buy shoes from that store under *any* conditions.

The service can be made better, of course. The mobile operator could easily filter out various customer groups. For example if the sale shoes are predominantly men's shoes and only a few items of women's shoes, then the targeted ads could be sent only to men this time. Or time can be used putting distance between the last purchase, sending the ads only to those who made a purchase four or more months ago. By focusing even more and targeting the ad to perhaps only 500 people, the store gets a much higher advertisement-to-sale ratio than by sending location-based-spam into the neighbourhood.

And of course, the store gets immediately measurable data on which coupons were redeemed, and can measure how successful this campaign was.

To encourage service take up operators need to create the customer database mobile ad service and sell it to the local merchants. They will need early examples to prove the case for this kind of advertising detailing the benefits to the retailer. When built with care and forethought, the whole mobile advertisement opportunity can build the customer-relationship for the merchant into a true long-term and very close relationship. The merchant will know what shoe sizes, colours, styles, any given customer likes, etc. If the subsequent ad campaigns build upon the previous customer data, then it becomes a virtuous circle, where each message gets closer and closer to the true preferences of the customer, and the customer's purchasing behaviour will evolve more and more to that merchant, which in turn will focus its product offering ever more to its loyal customers.

There is the other direction to this same opportunity. It is the operator-merchant relationship. The first ad campaign will be an experiment, but as the relationship improves, so too will the merchant and operator become closer intertwined. Soon the merchant cannot imagine leaving the operator because of all the customer relationship data being imbedded in the operator's database. But the operator will similarly have to become ever more aware of its customer needs to serve this shoe merchant ever better. Another virtuous circle.

7.7 Forwarding Ads and coupons

A particularly strong benefit of UMTS is the community nature of mobile communication as we saw in more detail in the Me chapter. This same community power can be harnessed in mAd. We associate ourselves with people of similar interests, tastes, hobbies. If an advertiser finds us, and assuming the advertiser targeted us accurately to begin with, we can then act as the agent for the advertiser, and find more people of similar interest. For example if I am identified as a fan of Tom Clancy style books, and I get a coupon with 10% off the latest Clancy novel, I would like to forward that coupon to several of my friends. The mobile operator will need to construct the message tracking system to be intelligent enough to know which coupons are forwarded, to whom, and how many times.

There also needs to be a cost associated with that forwarded coupon. The cost needs to be defined beforehand, be controlled, and paid for by the advertiser (e.g. retailer or merchant). The rules about forwarding the coupon need to be clearly indicated in the ad itself, so that users will know if this is a coupon that can be forwarded – and if there are limits such as "this coupon can be forwarded to five different people. If you would like to send more coupons, please contact us by clicking on the click-to-talk button..." The retailer will not want to pay for the wanton delivery of millions of 'free' forwarded coupons, but equally, if the targeting is accurate and yields direct action (usually purchases), then the retailer will *want* to pay for all forwarding.

Who pays

Of course in all coupons, the advertiser should allow the initial recipient to forward the coupon at his own cost, for example sending it onto a friend as an SMS text message. That is distinctly different from the case where the initial recipient picks a friend and forwards the message at the advertiser's cost. The advertiser needs to understand the distinction between these two actions. If I forward an ad at my expense via SMS text messaging, it is like I bought a newspaper, noticed a coupon and clipped it out (or photocopied it) and gave it to my friend. It was my cost to deliver the final ad to its recipient, and the recipient did not receive the *whole newspaper*.

If the advertiser pays for the transmission of the ad and allows me to send it to any number of people, it is not like cutting coupons, it is like a private reprint of the whole periodical. It would be as if the advertiser gives me the right to reprint the whole newspaper where the ad was, and publish it, and send it to my selected readership – to get a 'second edition' reprint of the full content. On a daily newspaper printed on paper, this is of course utterly impossible, but in UMTS it will become possible as electronic content and is quite possible to become the most common means of mass-delivery of advertisements.

Community forwarding

The issue here is identifying the target audience and delivering the ad. In a targeted UMTS ad, the real power lies in profiling. Using current data-mining tools, it is possible to delve into databases and accurately find the right target audience members. Today to do that from various incompatible and contradictory databases would be remarkably expensive. As UMTS operators collect data on their customers and these customer migrate their behaviour more to the UMTS environment, the databases will build naturally. In 10 years it will be a trivial matter to find all 10,000 fans of the Rolling Stones in Stockholm, Sweden!

Community forwarding is a short-cut to getting that data now. All the advertiser needs is a 'seed' target of perhaps 100 definite fans, and let community forwarding take care of the rest. If the forwarded message has an enticement for the recipient to contact the original advertiser for example by joining in a trivia game then the recipients are then validated and 'caught' into the database. Of course these are then encouraged to forward to their friends and so forth.

Eventually the same person will get ever more referrals from many sources so the resulting data will need to be cleaned and duplicates purged. The process will eventually also automatically dilute, as someone who is only a marginal fan, sends to another who is even less of a fan, etc., so that new referrals are no longer eager to sign up. By some trial-and-error the operator will be able to design community forwarding mechanisms which will accurately find anywhere from 80 to 90% of the real target audience, at least on such areas where there are vocal fans willing to be identified.

Yes to cumulating – no to tromboning

The advertiser and operator will need to control how accumulation will be allowed. Typically if one 'basic' coupon gives a discount of 15% on a late seat at the theatre, then two coupons should not give 30% off, unless for some reason the advertiser wants to promote accumulation of coupons.

But one must keep in mind that receiving coupons, and getting discounts are two different things. If there is an earned benefit which converts to discounts, and this is related to coupons, there is no reason why coupons could not accumulate. The typical use is to reward the person for forwarding with a discount for having found a suitable target person who has not yet received the coupon. There is no reason why multiple coupons could not accumulate for the very same service, if the target users 'earns' the benefit.

For example with the Tom Clancy book, if for every delivered coupon, where the target buys the book with the forwarded coupon, I get 10% off, then of course I am acting as a sales agent.... The more completed sales I get, the more I can be rewarded, and there is no reason why if I get 10 friends to buy the book that I could not then get the book for free. I could even sign up 20 buyers and get *another* Clancy book *also for free*. But this depends on how the forwarding and rewarding rules would be set up. If I cannot get more than 10% off, then I am quite unlikely to bother to forward it to more than one of my friends, and I will then ask *him* to forward to my other friend, etc.

The nearest analogy to community spreading of electronic coupons, is that of third party leads delivered in sales. Here a third party, for example a consultant or technician, might get paid a sum if he provides a lead which turns into a sale. There is no inherent absolute limit to how much could be earned by the leads.

The key to accumulated benefits from forwarding is that the operator will have to have accurate information on forwarded coupons. This must cover all contingencies, such as forwards into other networks (and back) and through various media and devices.

11 Vignettes from a 3G Future:

Movie Ticket On Mobile Phone Screen

Now there is no need to wait in different lines at the movie theater. It used to be that I had to wait in line to buy the ticket to the movie, and if I didn't make it early enough to the movie theatre, I would get bad seats. If I ordered tickets from the fixed Internet, I still had to show up an hour before the start of the movie, and often stand in line at a special counter to pick up my ticket. But now I can surf the movie listings and theater seats for free on my mobile phone and review trailers of the movies that are playing and of course select and pay for my seats with my mobile phone. Best of all I don't have to stand in line when I show up at the movie theatre, I just show my UMTS phone to the doorman who validates my ticket electronically.

The mobile purchase of movie tickets is likely to be one of the early popular services on UMTS. Movies appeal to the youth segment. Movie producers and distributors will be eager to place their trailers one minute advertisements of their movies onto the UMTS screens. They are likely to be willing to provide the connection for free just to promote the movies. The movie houses are eager to automate ticket sales and also minimise needs to man ticket counters. The savings from these personnel cuts can be passed onto covering the airtime fees of handling movie listing requests and movie ticket purchases. The savings in ticket processing costs will probably totally cover the costs of the air time and movie theaters and UMTS operators can make a very profitable setup out of providing mobile movie tickets.

The big danger is tromboning. This means back-and-forth repeated traffic. For example if I received the initial Clancy coupon with 10% off. I send the Clancy 10% discount coupon to my friend Timo, and then he sends it back to me, that I will not now get 20% off, else I will again send it to him, he back to me, and after 10 iterations, we both have the book for free. Tromboning – which could easily be triangle-tromboning i.e. I send to Timo, he sends to Joe, and Joe sends to me, and we do this carousel 10 times – can easily totally destroy the opportunity in forwarded coupons, and of course the danger is that content will be 'stolen' or received for free.

mAd bonus points

A variation of the mobile ad is the 'free headline, **click to view**' ad where any viewing of an ad would then result in a small gain to the person viewing the ad. This could be similar to airline frequent flier plans. The more you view the more you gain points. But viewing the same ad twice will not get you more points. The advertiser uses its profiling to send the links, and adjusts those profiles as it observes the resulting behaviour. It would be very easy from a technical tracking point-of-view to eliminate those who only click for the money without ever making purchases.

The customer on the other hand, always get to choose which – if any – ads they want to see, and in all cases when he does, he gains more points. The points could be redeemed in a variety of ways, from minutes of airtime with the mobile operator, to free trials of new services, to goods and services from the company advertising, to even real frequent flier miles on your favourite airline.

Frequent forwarder points

Just as airlines award frequent fliers, so too should mobile advertisers and mobile operators reward frequent forwarders of ads. The benefit could be in the form of points that can be redeemed with the advertiser or the mobile operator, or could even be convertible to an airline frequent flier plan, etc. The issue is that the person who received the ad, should be given motivation to forward the ads to others who might like them. Here too, there should be a control, such as awarding the frequent forwarding points only if the receiving person does click to the advertiser's website, or makes a purchase, or redeems the coupon, or whatever is the action needed. There is also the danger that anyone who has many good friends ends up getting the same forwarded ad 56 times. The operator has to then ensure that each ad can only be redeemed once by any customer, and that tromboning and other abuse is not promoted.

7.8 Free trials

The fixed Internet brought about a significant development in the way digital content is advertised by the way of free trials or 'tasters'. This is the case with some of the Internet games where a limited version was available for free with the option to purchase the full version once the 'addiction' had set in. The same format exists with the adult entertainment industries where the free trial model has been implemented to very good effect, such as viewing part of an image before paying to get access to the actual site, etc. The digital domain brings many particular variations on what can be done with free trials. These range from partially blocked or poor resolution content – you can see enough to know what it is, but not enough to be able to consume it to your full satisfaction. This is very common now for example with consultants' and industry reports, where the table of contents is released to show what is in the report, but the content has to be paid for.

This approach can be used with news feeds for instance. Headline news pages can be sent to subscribers of a low cost or free news feed service. If the full story is accessed by clicking on the headline there is a cost for that content.

Limited time free

Another free trial is variations around the time limit, where there can be a calendar time that will cause software or access to expire, or an amount of times that one can play the game, or use the software, until the right to trial it expires. There were some ways to manipulate these parameters when trial software was released on diskettes and consumed on stand-alone PC's, where the user could reset the system date to access software that was supposed to have expired. With operator-network controlled time, there is no such easy way around the time dimension for the habitual free usage customer. Equally the operator can easily track the number of times the software, content or application is used, to keep accurate count of each free trial and screen out the hyenas. It could even be built to be so friendly, that if one samples a game for a very short time, it is not 'counted' and the network could inform the user that since you were on line for such a short time, we will not count this as one of your 10 free trials. But in such cases as well, the system must be on alert for abuse, and make sure that the user will not abuse the short time good will rule.

Cannot save

Another variation is to allow limited convenience, such as allowing single use of software which cannot save, or does not have the filters for convenience, or other limits to its utility.

First level free

Yet another application comes from the innovations in on line video games where the 'first level' is free. This allows full use of one level of the game (or other software) but further levels have to be purchased. This also has many varieties with news content on the fixed Internet, where headline content or daily content is free, but deeper analysis or archives, etc., are billable services.

All of these are again forms of marketing through free trials, where the owner of the content gives a free trial or test or partial enjoyment of the subject to be consumed, and the buyer will need to authorise payment on the full content.

Through the advent of micro-payments, the possibilities of free trials and near-free trials is almost endless. A news provider can give headlines for free and charge extra for the content. In the case of content such as today's news stories that are old next week these can have their price gradually diminish over time as the latest news reports come out, still generating revenue from information which is quickly at the end of its utility. Trials can expand sequentially to more, so if a customer plays a trial game for free, the next level could cost 20 cents, and the following level 50 cents and the next one cost one dollar. Over time those who bought the full game could end up paying for example 20 dollars but since the cost is spread over weeks or months the perceived cost is lower than the actual cost. Even with various software license/payment/free trial mechanisms on the fixed Internet today, this kind of stepped approach is not possible as the credit card companies' fees to handle the small payments would eliminate any money to the content provider.

7.9 AdPay (PromoPay)

The success of prepaid subscriptions in GSM (Global System for Mobile communications) has brought with it one clear restriction, namely that the operator does not know the user. They remain in effect anonymous. This is fine when the main service is voice or basic SMS but in the UMTS service environment operators need to build up a user profile so they can customise their services and target them towards specific user groups and segments. Many of the prepaid users have low to medium usage and are a clear target audience for promotionally subsidised UMTS services, including voice. Many free SMS information services are already working today where users accept advertisements or promotions to their GSM or WAP (Wireless Application Protocol) terminal in return for SMS or WAP based content. In the UMTS service provisioning environment this brings the opportunity to combine the advantages of prepaid subscription with push promotions that subsidise the usage of new services to create an AdPay or PromoPay kind of service.

For this to work the UMTS user would have to register onto the UMTS operators network, via a mobile portal for instance so that they could accrue Promo Points to their account. The Promo Points system would work very much like today's frequent flier services where the more you fly the more you earn. In AdPay the more adverts you view or coupons you redeem, the more points you collect. Promo Points can then be used in a variety of ways; to credit your mobile account to subsidise usage or to purchase items via mCommerce like cinema tickets.

The amount of points earned could be variable depending on the total end to end transaction. If a UMTS AdPay subscriber only accepts a promotional message there would be a small award of points. If on the other hand an mCommerce transaction resulted from the promotion then a higher award of points would be made. Cashing in mobile coupons from the terminal could also result in Promo Points being credited to the AdPay users account. This kind of solution is not possible in the fixed Internet today because there is no end to end management and control of the transaction. The web page owner has no visibility if the banner click-though resulted in a sale and the advertiser has no way of knowing how and from where the purchaser arrived at the site. In the UMTS network solution there is this control since the UMTS operator can manage and monitor the service and charge for the transaction.

It would be necessary to ensure that users did not intentionally view tens or hundreds of promotional messages a day just to build up points. This kind of abuse already happens in the fixed Internet where subscribers to a service that pays small amounts of money for every banner click-through, have created programs that click-though web site banners 24 h a day, 7 days a week and generate reasonable sums of money for their sleeping owners.

Since people have a limit to the amount of information they can absorb there would have to be a limit on the number of mobile promotions that any one user would receive in a single day. This service is an ideal way to remove the anonymity of current prepaid users without increasing the credit risk advantage that prepaid was designed to enable. If an AdPay user did not have enough credits then the cost would be deducted from their prepaid card.

Apart from the current and future UMTS prepaid users this kind of loyalty scheme is an ideal way to kick start mCommerce for all users since every purchase or transaction that generates revenue for the UMTS operator would generate Promo Points for the user. This model is not new, it has been working for some time and has the same potential in the UMTS business model as in any other business providing it is positioned and marketed in the right way. So who will be the first Platinum or Gold UMTS subscribers?

In any case, advertising agencies will tell you that people can only absorb a finite amount of new information and the more promotional noise in the environment the less chance the message has of getting through and being remembered. Initially advertisements on mobile phones will have novelty on their side and will probably attract some of the most creative advertisement creation talent as it is the newest medium. Later on when user profiling can help the UMTS operator offer tailored segments based on previous usage patterns to the advertising agencies there will be more sophisticated Mobile Internet promotional services on offer.

Acceptance comes gradually

What Mobile Advertisers and operators need to remember is that the mobile device is often the single most important communication tool or in some cases the most important electronic device in a person's life. It is within arm's reach 24 h a day. The users will not thank you for filling their lives through their personal UMTS phone with junk and garbage. There has to be a different approach taken. The underlying economics will not be enough to win acceptance. Most people do not consider that the episode of their favourite TV programme would cost 10 times the amount they are paying if there were no advertisement breaks every 20–30 min. The newspaper or their favourite magazine have always been priced low enough to be purchased every day, week or month. Would they pay for the newspaper or magazine if it reflected the true price without the advertising subsidy. Unlikely.

The same approach is needed with mAd. If it is there from the start, subsidising the cost of calls, data access, services, music downloading, video streaming, sponsoring games, it will become the norm. We could even envisage that companies would accept this practice as well since it will reduce their communications costs, just like corporations purchase the same Time Magazine and Economist and Fortune as consumers, and those all contain advertising. It is difficult to imagine that early on the corporate world would embrace advertised content, but in a cost-conscious environment, there will be companies which will be interested. If, for example the mAd subsidised UMTS service was 20–30% below the non subsidised service perhaps many Chief Financial Officers of large corporate companies with thousands of UMTS phones could find the cost saving substantial enough to make it compulsory. After all, it is the company that is providing the phone and paying for the calls. The old rules are out of the window, welcome to the new world.

The winner takes it all

The opportunities for mAd and mCommerce are huge and they have much more real potential than just hope or hype. The returns will be delivering better content and value to the users and the industries involved.

Mastering the market opportunities will require careful planning, strategic choices of the right partners and the right vehicles and a fair amount of costs for setting up the channels. For mobile service providers success in promotions and sponsoring is all about being attractive to the advertisers. Similarly the eCommerce and bricks-and-mortar merchants will want strong partners to help them get into mCommerce. Progressive branding and unique content will drive the growth in these industries, whereas average concepts will face hard time in gaining the attention of and success in the UMTS marketplace. The ability to serve the advertisers and merchants in the form of customer information, measurements, profiles and location information is crucial. Becoming a trusted partner Money community should therefore be a strategic priority for the mobile service providers.

7.10 Show me the money

This chapter has looked at some services which have a strong benefit on the Money aspect of the 5 M's of Services for UMTS. The services were not an exhaustive listing, and described mainly only to provide a deeper understanding of the types of UMTS services that are expected to be created. Each of the described service would have several of the other 4 M's as a strong attribute as well. In designing UMTS services the operators will need to experiment and be creative.

Potential services that can capitalise on the Money aspect number in the thousands – theoretically every service will need to have a Money dimension. And operators should not worry too much early on in trying to get it exactly right. With a new technology it is very difficult for users to give exact guidance on what they really want. Probably if asked in 1945, most radio listeners would not have imagined being addicted to TV game shows or late night TV. An appropriate thought would be Mad Magazine's Alfred E Neuman when he said: "Most of us don't know exactly what we want, but we're pretty sure we don't have it."

8

'Any sufficiently advanced technology is indistinguishable from magic.'
Arthur C Clarke

Services to Address Machine Needs:

Empowering Gadgets and Devices

Joe Barrett and *Tomi T Ahonen*

The last major area when examining the 5 M's (Movement Moment Me Money Machines) is Machines, allowing for devices to perform activities and to communicate. This type of listing will probably be endless, but some early obvious areas are automobile telematics, home appliances, metering devices both fixed and mobile, robotics and voice activated automation services.

The services described in this chapter have a high benefit on the Machines attribute. There are hundreds of such services and we have a chance only to explore a few of them. But to illustrate by way of example some of the obvious uses of Machines type services, they include:

Business – control of devices, resources	Remote metering
Car safety	Remote monitoring systems
Click to talk	Robotics control
Commercial vehicle telematics	Software upgrades to automobiles and gadgets
Home appliance control (vignette)	Software upgrades to computers, game consoles
Identity verification	Software upgrades to mobile phone
In-car entertainment	Telehealth monitoring patients
In-car security	Traffic assistance
Live camera feed	Travel guides and maps
Music composition	Virtual home environment
Personal assistant search engine	Voice IP (Internet Protocol) telephony
Personal information in car (vignette)	Wireless office solutions
Promotions via search engines	
Remote control systems	

The mobile phone started its life as a car phone for the VIP executives. These were so big and cumbersome that with their battery, the phones were sizes of briefcases and weighed too much to be really classified as mobile phones. In the early days there was a very low proportion of hand held phones compared to those in cars. Early mobile phones were seen as snobbish gadgets and some who had them were known to drive around holding the phone without a phone call, just to show off the fact that they had a car phone. We've come a long way since that time.

The mobile phone of course grew smaller as technology developments were made in microelectronics and battery technology and soon became truly portable. Today it is not longer seen as a 'car phone', more a personal mobile device for voice with a growing data services component. The car was a natural environment for the new mobile phone (or cellular phone in American parlance) and car-kits were sold so that users could connect the phone to the car to use the electricity and antenna. Now with the advent of automobile telematics, we are coming full circle. The mobile 'phone' is using the same technology – UMTS (Universal Mobile Telecommunications System) radio waves – but providing totally new services, where a voice call may be an unnecessary extra for many drivers. Beyond car telematics, there are the various other uses of automation, robotics and gadget connectivity, where no person is necessarily involved in the communications traffic. Let us begin, however, with a familiar family situation.

8.1 In-car telematics

Any parent with small children would pay a large sum of money to quell the fight in the back seat or to answer the question "Are we there yet dad?" with a response like "Check the map on your console and see where we are." Even better then if the question never comes at all because the children are playing Mobile Internet enabled games or chatting with their friends via the in-car UMTS system.

Our need to be in touch doesn't stop when we get into our cars. And being in touch doesn't mean anymore just voice: users want their e-mails, news and other information as well – and what would better than checking e-mails when sitting dead still in a traffic jam or downloading new games for the kids on the backseat when taking that 5 h drive to see the grandparents. The kids may end up posing the opposite question, when in the UMTS future they ask "are we there yet" – they may be fearing that the car trip will end before they can complete their ongoing game session.

In-car telematics is widely used as a buzzword for applications and services targeted for people travelling in their cars, ranging from roadside assistance and e-mail access to downloading games. The in-car telematics market has existed already for several years through CD (Compact Disk) based **navigation systems** but they haven't really been a thing one could call a big hit. Volumes have been low as end users perception of the value-price equation has been on the negative side and virtually no company has been able to make good profits in this area. One clear limitation is that the information on the CD is out of date as soon as it is pressed. Low end systems rely only on this static data without taking traffic or road works into account. Of course high-end systems exist which have dynamic parts that adopt and update the information on a continual basis but they have not yet become a mass market product.

Route planning and traffic information

Today if a user is surfing the Internet and look for services related to travel and cars, they would come up with different Internet sites including SMS (Short Message Service) and WAP (Wireless Application Protocol) based services for route planning and traffic information. There are lots of small companies working with WAP solutions but when looking for bigger players, it is hard to find too many of them outside of the UK, France or Germany. Traffic Master[1] in the UK is one such company that provides updated information of traffic congestion for major highways and is now looking to expand into Europe and America. This system requires the purchase of a dedicated terminal that can only be used for traffic information provided by this one content provider (Figure 8.1).

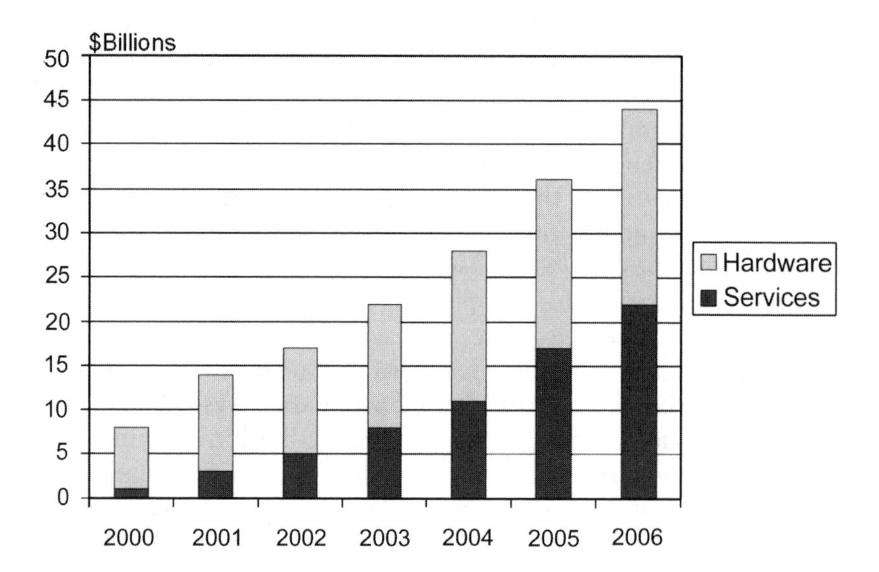

Figure 8.1 World Telematics Market Value. Source: Motorola estimates.

The future of in-car telematics

With the emergence of packet based wireless technologies, a market uptake for in-car telematics business can be expected, that would

[1] http://www.trafficmaster.co.uk.

benefit terminal, application and network suppliers as well as service providers who would actually be providing the service. Naturally the end user has to benefit from the new technology as well, otherwise there wouldn't be any sense in this equation. The force behind this expected rapid revenue growth of in-car services is increased bandwidth, together with advanced application technologies. These have the potential to introduce new, rich services for the car environment that not only provide all the necessary information and entertainment but will also monitor the car's performance and personal security. All this can be enhanced with easy of use terminal, especially designed for in-car usage via a specific UI (User Interface).

Car manufacturers are actively pushing the concept of integrating modern communications into the car. These days every major autoshow has concept cars from different manufacturers which enable browsing of the Internet, downloading the latest videos for the children to be played on the backseat video screen and trading or checking stock portfolios while stuck in yet another traffic jam. The automobile manufacturers are racing to deliver the **smartest car** and to have their fleets compatible with this vision of the future. As can be expected, Japan is way ahead in this area with for example live cameras at various traffic choke points in Tokyo and other major cities, and cars which can access the camera so that the driver can see if the traffic is bad on a given route. The Americans are also very far ahead in this vision riding on the IT (Information Technology) leadership of silicon valley. The United States is the largest car market attracting the innovation and paying for development is eased by cars tending to be larger and replaced more often. In Europe the development tends to be with up-scale brands and the high-end of the model ranges.

From the car manufacturing point of view mobile communications has become a new area to gain a competitive edge, customer loyalty and brand awareness. The dilemma many manufacturers face is that customer loyalty is very hard to keep if the only time they visit the dealer is the once a year service stop. When it comes to advanced in-car communications solutions the service provisioning business is something that the car manufacturers are looking at very closely in order to build closer relationships with their customers, to offer better and new services and to position their brand.

UMTS operators as well as other service providers have a keen interest in how the car environment can be integrated into the Mobile Internet. When defining their service strategies for the next few years the car environment is a natural extension to their offering to serve their customers better in all environments and situations. The UMTS era is about services and new sources of revenue that require new ways of thinking. The in-car segment is one area that will to be taken into account and could prove to be very lucrative.

Who owns the user?

A car manufacturer as a service provider? The idea is not that far fetched as it first may sound. Cars are old business with established players, high value networks, established distribution channels but with low industry growth figures. It is obvious that in the near future people will be using more and more Mobile Internet services not only at home or at work through the personal UMTS terminal but also in their cars. The big question is, will the same service provider have solutions for all those environments and have the ability and knowledge to serve all the users specific needs? It seems to be that there are car manufacturers who say that once you buy their car, they can also provide your wireless services. Naturally the operators say the same thing, "If you subscribe to my service we will take care of all your needs, including in the car" And then there are also the service providers that don't have a license to operate wireless networks but provide their services transparently to your mobile phone via the network provider from where you purchase your SIM (Subscriber Identity Module) card.

One of the main questions for the UMTS era will be, who actually owns the customer? It is not even clear if there will be any of the identified and even unidentified organisation that will be able to stake that claim. Today the company that has provided the SIM card and mobile phone usually also built the mobile network holding a very strong market position. But in the future the rules can change and especially the car industry will not take a back seat in the forthcoming battle. As services will be the key issue, will the same companies that provide services for mobile phones be doing the same thing for the in-car environment? Maybe users will get everything needed in the car from the same company that sold the car or maybe there are different portals from different service providers that deliver specific end user services depending of what device is used to access the Mobile Internet. Even in this in-car services business many companies are claiming publicly that they will be the leaders, the only real alternative supplier with the ideal position to deliver the value required by the driver.

At the end of the day the majority of players are expected to migrate towards cooperation and partnerships, as the way to move ahead and create the best solutions that is based on each partners' core competencies. In this context, car manufacturers and traditional service providers would partner and combine their expertise to deliver the best in class service whether in, or outside the vehicle. So without having a direct answer to the question presented in the beginning it can be said that which ever model will win, the best positions in this race over consumers in their cars are occupied by operators and car manufacturers, who both have their large existing customer base, knowledge of their products and future roadmaps as well as the usage patterns and habits of their customers.

It is easy to understand the car manufacturers and their interest to expand their business into service provisioning. Suddenly this new technology shift offer new ways to enter the fast moving wireless business and capture new revenue flows. But car manufacturers have obstacles they will face, not least being their inexperience in delivering wireless services to their customers. Car manufacturers will need strong partners to succeed in the services business from the very beginning. From the service providers' and operators' point of view the in-car environment and services is just an extension to their current service offering. Although this is can be major business from a revenue point of view, the business fundamentals and the problems that the current mobile operators face in moving from network operators with a few services to a total UMTS service provider will also be experienced by the car manufactures and dealers as they enter what is in effect a totally new area for them.

One of the main open questions that will affect the success of car manufacturers and wireless operators is which terminals end users will actually use when they are sitting in their cars. Is the large and powerful in-car terminal the chosen one or will future UMTS terminals be good enough to be used for accessing services while driving a vehicle. If the integrated in-car terminal becomes the de facto solution, car manufacturers have a strong position as they can install the terminals at the production line and perhaps bundle it to the basic car pricing. In this scenario all other players must provide after market solutions where the terminal is installed in the car after purchase. Naturally they can also cooperate with the car manufacturer. But if the mass market mobile phones remain as the preferred terminal also within the car, wireless carriers have a strong position. We believe both models will exist and find their niches, not unlike car radios today. There are built-in radios especially at the low end, and major brand car entertainment systems at the higher end. With UMTS services probably several varieties will exist to suit the driver's – and passengers' – preferences.

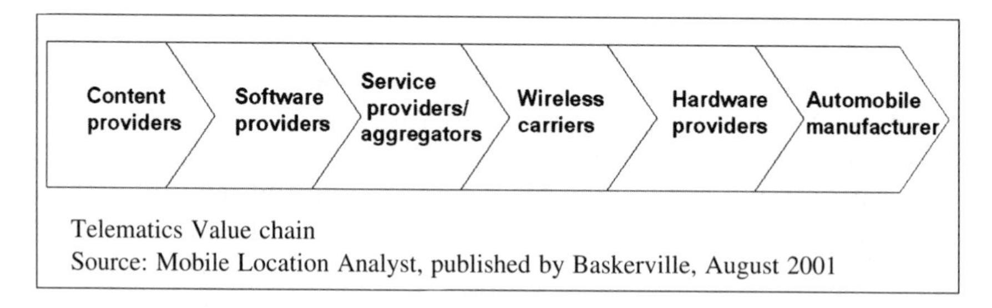

Telematics Value chain
Source: Mobile Location Analyst, published by Baskerville, August 2001

In-car communication and information

What are the services we want to use when on the road? For road-warriors as well as busy executives both men and women, the basic communication and information methods used also at the office are important. Voice, e-mail, calendar, Internet, Intranet. During weekends and holidays the same people load up the car with spouse and children and head somewhere for a little rest and relaxation. At that time probably everyone who has a family would agree on spending a few dollars to keep the backseat people busy with new games, electronic books or something similar.

In-car safety

A major category of car services relate to safety in accidents and mechanical breakdown. When the car's airbag goes off the instant safety need is to have the rescue organisations notified about incident, notification of the exact location and potentially also speed and heading at that time. It should be possible for the emergency services to be able to contact the in-car system and check the situation with the vehicle and also make voice and visual contact with the occupants. In the unlikely event that the car breaks down, the need is for the fasted possible help.

Car diagnostics and repairs service

In an accident or break-down situation, remote analysis information can be sent to the service support company. This would allow the mechanic to be warned of what the potential problem is and allow him to bring the right parts, reschedule work to the right kind of mechanic and if necessary arrange more than just roadside maintenance. In some cased remote correction may be possible as more and more software is being integrated into vehicles. Even better solution is if the car could provide proactive information possibly based on remote analysis from the service company that there is something going wrong. Advice of what kind of problem is brewing, where to find service and how urgent the situation is could then be communicated direct to the driver.

Personal security

Another area of ample service opportunity is security. This varies greatly between countries, and even regions within countries. Most societies have to worry about more traditional car theft, and by traditional we mean that you return to where you parked your car only to find it has been stolen. In some societies criminals actually hijack cars from their drivers, for example when the car is stopped at a traffic light, by walking up to it and pulling a gun. In America this crime has a name: car-jacking. In some countries people have to worry about criminals committing crimes on motorists ranging from robbing them to rape and murder. In other countries kidnapping is a very real problem. In yet other countries people have to fear car bombs.

The UMTS services will allow many security oriented features to be built into different kinds of devices and here the severity of the threat and money are probably the only limits. Simple services already exist which allow a car owner to **remotely start up the car**. This not only has a security benefit, but also that of preparing the car for harsh conditions – turning on the heating in cold climates, or the air conditioning in hot climates. Services can be built to **view the car and/or parking lot**. Remote control **ignition turn-off** can be created which would stop the car if it is stolen. This type of service needs to be gradual, for example slowly removing the power from the gas pedal, otherwise if the car is suddenly stopped, there could be a serious accident by which another motorist drives into the stolen car.

There is no limit to what could be created. A **panic button** could be built into the car to provide a feeling of safety for the driver, and to deter potential criminals. The panic button would provide rapid contact to the emergency service centre and set up an instant video call between the car and the police.

Pay parking & fines

As has been seen in early chapters users will become accustomed to paying small fees via the UMTS terminal. This will be extended to parking and maybe even to parking fines. There is no end with these examples. There will be differences between the North American and European markets in terms of which services will be the most popular, the emphasis in the American market is very much on personal security although this is also important in all areas.

In all probability many people will consider that services in a car environment will bring value and are essentially a good idea. The wireless services on mobile phones have become hugely popular and in the vehicle the screen size will not be necessarily limited by the need to fit the device into a small pocket.

In-car entertainment systems

Cars, especially the rear and passenger seats will become more like aeroplanes with large screens, personal headsets and keyboards. With the diminishing costs of flat-screen displays and ample power from the car's engine, the services for the passengers can mimic those that we see on the fixed Internet today. Any kind of games, entertainment, information and communication can be enabled for the passengers. Some specific car-entertainment solutions will of course emerge, relating to car travel, the journey taken, real traffic conditions etc. Perhaps for the backseat drivers a virtual reality game where the passenger could **drive the virtual copy of the family car**, on real traffic conditions, via the passenger's preferred route, to see if the backseat driver could have navigated to the destination faster than the driver.

For the driver it is highly dangerous to be participating in a video call while driving and many countries already ban the use of video or TV screens that can be seen by the driver. The systems have to be designed with driver and passenger safety as the most important feature. The police are unlikely to be happy with you trying to break the Tetris world record while driving at 80 km per hour on even an empty road. Many countries have banned the use of mobile phones while driving without hands-free equipment and this is expected to be extended to new cars in UMTS services.

Voice recognition in car

This dilemma is, at least partly, solvable through implementing voice recognition technology. Instead of pushing buttons while driving the driver just says the word 'e-mail' and the inbox opened and a voice-menu will offer to read e-mails, take dictation, forward e-mails, etc. The in-car unit can read e-mails to the driver enabling the eyes to be kept firmly on the road and not on any screen. This doesn't help with the Tetris problem, but is a major issue with the basic day to day services. And like with seat belt or hands-free, the driver is obliged by law to use them. Break the law and get caught? There will be a fine.

The money side of car telematics

After reviewing briefly the basic market situation, players and potential services, the question of volumes and potential revenues remains. As always, predicting the future is not one of the easiest tasks but here are some figures where each reader can draw their own conclusions: The European and Northern American new cars market are roughly on the same level in terms of units sold, around 16–17 million passenger cars a year. Europe, United Kingdom, Germany, France and Italy make up a very large part of the total sales so when planning to do something for the in-car environment, those markets should be the first ones on the hit list.

The penetration of in-car terminals in new cars is a major factor when calculating the overall market size. Many sources predict that by 2005–2006, roughly 50% of new cars are equipped with some form of in-car terminals (which may be nothing else than a UMTS terminal). The main consideration is if consumers actually use these new UMTS services and what are they willing to pay? Looking at the solutions that are currently on the market and have paying customers, an area of 15–30 dollars per month could be achievable. A family's security is worth at least this much per month. However, it is unlikely that end users will pay this amount in addition to their other mobile spending. This figure includes a comprehensive package of core security, car and traffic related services as well as more general services that the customer pays for in any case. The pure in-car telematics might be perhaps maximum 50% of that figure. Operators will probably have to create a general in-car package at a much lower level and allow users to top this up with other services as needed.

So it seems to be obvious that there is market potential and there are varying figures offered from all consulting companies on the size of this market in the coming years. Whether these figures will ever be reality is up to technological developments, standardisation as well as the development of innovative and informative services where price and perceived value meet the customers expectations. There are major obstacles still to overcome, so any market figures shouldn't be taken for granted. For example, in many countries traffic information is available through reports on radio, so in order for end users to pay for the same basic service there really has to be clear value added over and above what they receive today. Location can play an important role here and this is expected to become the main differentiator in these kind of service evolutions.

Challenges

So by now readers could be exhilarated about the new applications that will be available in cars in a couple of years from now. However we need to keep our feet firmly on the ground. There are still issues to be solved before the forecasted market volumes can be reached. One major issue is lifecycles: when something is put in the car it can be there for the next 10 years. This is a very long time from a mobile communications technology point of view. The replacement market for mobile phones today accounts for around 50% of sales with some vendors and the average user changes their phone less than every 2 years. An analogy can be easily found from PC technology. It is virtually impossible to run applications made today in a PC manufactured 10 years ago. Whether this problem is overcome by developing upgradeable terminals, mainly memory and processor, although this is not completely future-proof solution, or by simply accepting the fact that the hardware has to be replaced every 5 years or so.

Alternatively the in car system is a hybrid solution. The communication access is always the users UMTS terminal so this is always of the latest or most up to date part of the in-car system. Each person in the vehicle then has their own access via their own personal device. The in-car system then becomes an extension of the users UMTS terminal where things like screen size, keyboards, sound systems and accessories are merely extensions used in the car environment. Specific in car software like the service program would then communicate to the service provider only when there is someone in the vehicle. If we think about this logically, there is possibly only one service that is needed when there is no one in the vehicle and that is car theft alert. A specific unit could provide this functionality and is less likely to become redundant over time.

Unfortunately for the car manufacturers' this is not the ideal solution since it is placing the power clearly into the UMTS operators domain. In any respect how and in what way all the players manage to integrate themselves into this particular chain still remains to be seen. What is vital is that during the jockeying for market position the parties involved do not take their eyes of the user and what customers actually want as well as what they are willing to pay for.

In this industry as in mobile communications, standards will also be major issue and that hasn't been so vital for current in-car solutions outside of the legal requirements and framework. For UMTS in-car services, the same service from the same service provider must work regardless of terminal or manufacturer or supplier. The applications themselves as well as the methods and the way they are delivered to the terminal must be based on the mainstream mobile technologies including UMTS. If the in-car solutions evolve to be something different, requiring major investments in system and software development, there is a great risk of going back to non interworking and low volume solutions.

However, the largest challenge is probably the business itself. Technology can be made to work, but if there are no value adding services or terminals at the right time at the right price, no one will buy them. That has been partially the problem so far with in-car telematics. The perceived value versus cost has been too low.

8.2 Remote metering

When discussing telematics we should not forget the various measurement and monitoring instruments and control of robotics and devices. The unglamorous remote monitoring and control solutions will form the majority of the actual 'population' of telematics devices. In fact we expect the machine population for UMTS subscriptions to vastly exceed that of the human mobile phone penetration. The most relevant matter to keep in mind with remote diagnostics, metering, measurement and control of devices is that the traffic is predictable and mostly extremely slight. Reading your gas meter or sending an alarm from a **home alarm system** which contacts your UMTS phone would only involve a short coded sequence of a few bits of raw data. Even when converted to a full UMTS transmission this information would easily fit within a transmission of a few kilobytes. The traffic load from a given remote metering device is totally meaningless in scale when compared with the megabytes of bandwidth consumed by humans and their multimedia voice and data services.

Reading utility meters

The key here is not the amount of traffic, but rather the value of the transmission to the IT solution and its replacement cost in alternate means. For example if a mechanical water meter or electricity meter needs to be read by a human, there is a large cost of manning the pool of technicians who do nothing but visit meters to record their readings. This can – and should – be automated and read remotely. There would be less error and by far less cost. Metering traffic by itself is not the value, the value is in the information being delivered. It requires a fresh new approach by the UMTS operator.

Portable remote meters

Remote meters can be built to be dedicated UMTS devices. The real benefit would need to be derived from their portability. The most obvious uses are in robotics and other remote-control devices, as well as temporary metering devices, such as a flood control water meter or an after-the-fire alarm – which could be left after the disaster is over, to make sure no hidden problem remains. In some cases there is no fixed telephone line, and the UMTS (or other cellular network) may be the most cost-effective means to provide datacoms connection to a remote site. This could be for example in weather metering at locations away from urban areas.

In most cases today metering devices could be fixed using the existing telecoms cables, running on POTS (Plain Old Telephone System) or more advanced telecoms wired connections such as ISDN (Integrated Services Digital Network) and ADSL (Asymmetric Digital Subscriber Line). Some of these connection solutions are however cost restrictive and the advantage that the UMTS operator can bring is a total cost saving in manpower and operational overheads. UMTS operators have a greater flexibility in pricing there service since the radio resources are more easily shared than fixed communications distribution systems.

12 Vignettes from a 3G Future:

My Car, My Secretary

I love my new car. It has the wonderful smart telecoms setup. One of the neatest features is that the service is combined with my e-mail. So now as I drive I can have my e-mails read to me. The voice that my car has is a wonderfully pleasant voice and now I can have any of my e-mails read to me while I drive. If I feel like returning to an e-mail or if I want to view its attachment later, I can do that of course. The whole service runs on voice promts, so I never need to touch anything, I just speak to my car to skip to the next e-mail, or re-read a passage.

While cars are likely to become ever more loaded with electronics and intelligence, one of the big factors is that a high proportion of business executives tend to spend time in their cars. They often try to make good use of this time by placing phone calls, etc. But one new aspect with emerging text-to-voice solutions and the integration of e-mail to such solutions, is that the e-mail which can arrive to the UMTS phone, can also be converted to text. While many places may be convenient for the UMTS phone to act as such as "reading secretatry" one of the most used ones is most likely going to be the automobile.

Reading fixed meters with UMTS terminal

While many remote metering devices themselves may be connected to the fixed telecoms network, they could be *read* via an UMTS terminal or device. For example a security solution could be fixed in the building, but the security guard could inspect the status via his UMTS terminal while walking around the building or from outside in a vehicle after an alarm had been triggered. Other metering devices might be in such remote locations where regular phone lines are not present, and UMTS would be a cost-effective means to provide wireless access. Here an example could be a weather instruments station in the middle of a forest. Another example is the portable metering device, such as the Mobile Speed Trap.

Mobile speed camera

Speed cameras are a growing area of interest for many police forces and governments around the world. The UK for example has been very keen on experimenting and trailing stationary and mobile speed detection systems. In some countries these systems are gaining approval based on the results of studies that accidents are more likely to be fatal if speeds are even slightly above the speed limit for a particular road[2]. The total cost to society of speeding and the resulting accidents is difficult to quantify exactly but does impact the emergency services, hospitals as well as other road users. Although this is an area of hot debate with many motoring organisations claiming persecution of the motorist the saving in human life can be justified. Some countries like Finland base speeding and motoring offence fines on yearly income. Imagine paying 10% of your monthly salary as a fine for speeding. The highest fine as of time of writing in Finland for speeding was 500,000 FIM or almost 90,000 dollars.

What then is the potential for UMTS speed cameras. Firstly UMTS speed cameras can be left unattended for long periods and do not need any fixed telecommunications link. Pictures can be sent automatically

[2] Kloeden CN, McLean AJ, Moore VM, and Ponte G (1997) Travelling Speed and the Risk of Crash Involvement, NHMRC Road Accident Research Unit, The University of Adelaide.

over the UMTS network to the monitoring station for instant processing. These devices can be run on batteries and/or supported by solar energy. If the battery is running low a signal can be sent to the monitoring station so that a fully charged battery can be installed. So a motorist exceeding the speed limit by a certain amount measured by radar would be photographed. The picture would then be sent to the police computer for processing thus reducing the need for expensive film.

Real time speeding ticket

The service can of course be made even more sinister. The speeding fine notice or court summons could be sent in the post within the hour, but the elegant mobile service would of course send the fine electronically to the UMTS screen in the car within a matter of minutes of the speeding offence being committed. Committing the offence and then being instantly informed of the fine is probably the most powerful way to reduce speeding. In fact the service could even contact the driver or car directly and give verbal notice via the in-car UMTS system informing the driver: "You have received a speeding ticket. Please slow down. More mobile speed cameras are situated on your route", or something to that effect. The results on speeding would probably be quite significant.

This type of service would have a very different traffic pattern from a human. The speed camera would be uploading constantly images of speeding cars, never needing to take a break to eat or sleep, and in that way would probably load the network more than the most avid web surfer could hope to match. Yet from the police force point of view, the small costs of the picture transmissions would probably be significantly less than the costs of manned speed traps or rushing after speeding cars.

8.3 Remote control

If the first step to communicating machines is to know what they can sense, the next step is to use the communication to manipulate the remote device. Here the opportunity will arise more from temporary robotic gadgets than purpose-built machines. For example, if a mining company wants to deploy a series of robotic digging devices and remotely control them, the mining company is likely to invest in a radio control solution which uses unlicensed radio frequencies and invest in the transmission equipment as well as custom-built, robust mining robot devices.

But if an amateur geologist happens to find an interesting small hole in a cave, and would like to explore it further, the geologist could create a temporary roving robot out of a child's remote control machine and have his own UMTS phone with its in-built video camera taped on the top of the tank. Lego, the Danish toy supplier are already offering remote controlled electronic developer kits and many of the enthusiasts and purchasers of these 'toys' are well into adulthood. Although this may be a niche market it should be fairly simple to re-use off the shelf UMTS remote modules as part of a remote toy with a camera and control it from some distance. Even from another country.

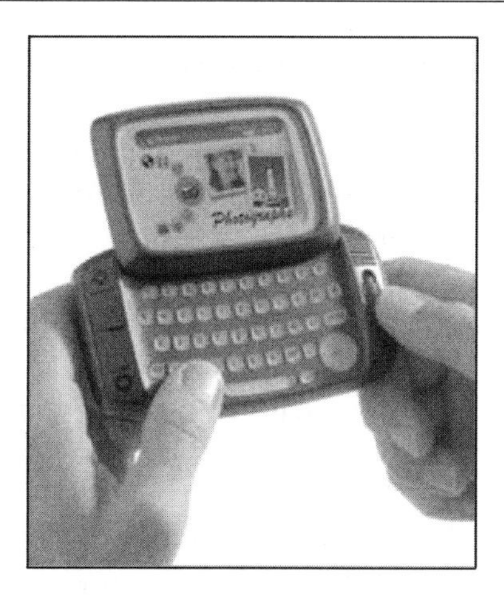

As the screen size of the terminal increases other options have to be found for the keypad. All phones will not be based on PDA-pen input methods and voice recognition although improving is not yet suitable for the mass market. One option to get around this problem is demonstrated with a variation on terminal design this time from Hiphop.

Similar simple uses are likely to spread to other niche product areas possibly driven by the development of Java applets that enable simple commands to be carried out. Examples could include:

A remote door camera or house cameras that can be viewed remotely via the UMTS terminal. This could be extended to an active video call being placed by the remote UMTS camera device if the door bell is rung while the occupants are out or if security lights come on in the garden or if the intruder alarm is activated.

Nursery cameras that enable parents to keep an eye on what is happening in class 5 min after they have left their distraught child in the care of these new teachers. Often the child is perfectly happy and playing with the other class members.

In-car security devices that activate a video call to the UMTS terminal when illegal entry happens. Remote deactivation if ignition and electronics and locking the perpetrators in the vehicle.

Remote activation of the video recorder so that an important episode of the x files is not missed.

Remote controlled toys such as cars, airplanes and boats enhanced with UMTS. When connected say to video glasses see the view from the remote controlled toy. Distance is not limited by short range radio as it is today.

Surveillance gear such as specialist equipment used by the police, guarding companies, etc., would be naturals to use UMTS video feeds and the ability to be programmed using IP standards and Java language.

Many of these applications will not be main stream but likely hobbies for some enthusiasts. The real likely uses will come from industrial control devices, which are mostly immobile, but situated in remote locations. Imagine the example of the remote reading of the electricity meter above. There would be only a trivial step beyond that solution, to include the 'on-off' switching functionality to the metering device. It has now transformed from only reading the electricity meter, to being also remotely able to turn on and turn off the electricity for example if the bill is not paid, or the person moves out of the apartment.

Industrial remote control

Taken a step further, we get the flow control of various processes. For example a water pipeline, gas pipeline, etc., has various points where its flow can be controlled. Some of those controls are very old, others are mechanical. There is no reason why these could not be digital, and the control be variable to decide how much water flow is passed from a given point to maintain water pressure or to raise and lower the water level of a water reservoir. In many cases these will be fixed line digital solutions, but equally, there will be many cases, where a UMTS solution can be cost-effective, even superior, to the cost of a fixed line to a given control point.

As we have seen with the space program, NASA builds the capability to **upload new software** to its space explorers. This kind of capability can be built into remote UMTS telematic devices so that new functionality can be added as for instance new Java applets are created for specific tasks.

Telehealth (telemedicine)

The healthcare market is seen as one area where telematics or remote monitoring of a patient's medical condition will be beneficial. This can extend right from a roadside emergency where a remote doctor is giving advice to staff at the scene of an accident to an out patient who, via a UMTS enabled device can have their medical condition almost constantly monitored. If set limits of say blood pressure or heart rate are exceeded a notification can be sent to medical staff. It would also be possible to monitor a patients ECG (Electrocardiograph) no matter where they are.

There are a growing number of Internet sites that offer various forms of medical advice even with on-line advice, some of it however is a little dubious. There are also a number of medical solutions that are using WAP to gain access to medical databases or for example to provide drug and pharmaceutical news to doctors.

The driver behind these kind of Telehealth services is the cost saving potential that could be realized say be letting patients home early with a remote monitoring device thus freeing up beds. The cost of providing health care is a huge burden on the state and any way of reducing costs will be actively investigated. This can be an opportunity for the UMTS operator to exploit what could be a new large revenue source.

Remote parking meter control

There are already a number of mobile parking meter solutions, mostly in European countries. Some are based on centralised database solutions where the car owner notifies a central database of the parking location, zone and car registration number. The charging is in place until the owner returns to the car and sends a message that the parking is ended. A bill is then added to the mobile phone account for the duration of the parking. Control of this system is by parking attendants who check that parked cars are currently registered on the database.

Alternative solutions are based around using the mobile phone as a mobile wallet. The parking ticket is given by the nearest machine to the parking zone on receipt of a message from the car owner. The cost is then added again to the mobile phone account so no money is deposited into the machine. The disadvantage of this system is that physical parking tickets are still needed.

There is scope for a dedicated UMTS parking terminal that could be used to display an electronic parking ticket. Validation could be by bar code that would be read through the windscreen or simply by the display showing the time of parking. Once the parking was over the charging details would be sent to the parking account and the mobile phone bill debited accordingly.

There are many similar options for paying for services like parking that sometimes require validation of purchase. Travel tickets, cinema tickets, entrance tickets. The GSM (Global System for Mobile communications) coke machine is often mentioned and there will be many other man to machine applications in the near future.

8.4 Shutting off the machines

Machine to machine and man to machine applications and services can provide regular revenue opportunities for the UMTS operator. In car services are an interesting area as such and they are also currently one of the hot topics for car manufacturers as well as wireless and non wireless service providers. There is a great deal of expectation and much hype for next generation solutions and only time will tell whether these expectations will be met. The UMTS era is about services and the car, health care, remote monitoring and mobile parking are all very good example of the new possibilities in terms of environment, technologies and applications that service providers are facing. New entrants, in the case of car manufacturers, may challenge the existing service providers in different vertical segments. Whether different vertical markets are served by one company with a broad portfolio or number of companies with more focused product portfolios – or by a combination of these two – is also a very interesting question.

The success of the market for Machines services is like with all services for UMTS, dependent on peoples' needs and how the products on the market respond to those needs. Whether users would like to have services for example into the car that provide useful information, entertainment and security is not in question; the market will materialise, it is just a matter of when. The 1 billion dollar question 'what's the price?' The world where machines will be empowered will be again a better place, as Isaac Asimov said: "I do not fear computers. I fear the lack of them."

9

'Change does not necessarily assure progress, but progress implacably requires change.'
Henry Steele Commage

Types of UMTS Services:
Categorising the Future

Joe Barrett, Ari Lehtoranta and *Tomi T Ahonen*

The 5 M's (Movement Moment Me Money Machines) are a useful way to explore attributes of UMTS (Universal Mobile Telecommunications System) services, but the 5 M's are not a tool for categorising services. As most UMTS services can be enhanced by each of the 5 M's, soon all mature services could fit in any of the 5 M's. As the UMTS environment will be introducing thousands of services, many of them radically new from existing services, there needs to be some ways to group services. This chapter will look at how operators, content providers and application developers can categorise services.

Readers may have noticed that many of the example services described in this book can be deployed on current technology networks. A valid question then arises on what are the types of services which are strictly UMTS services. Some of the services discussed in this book can be deployed on current second generation networks using SMS (Short Message Service) technology and many of them can be deployed using '2.5G' technologies such as GPRS

(General Packet Radio System). On UMTS networks however the services can gain from technological benefits like bandwidth and capacity increases that will bring cost-efficiencies as well as a richer user experience. Therefore it is worthwhile to compare the differences between types of services on UMTS and those services on current 2G and 2.5G networks.

9.1 Comparing 2G networks and UMTS

As technology approaches a discontinuity or new generation, it is natural to ask, "What are the differences between the existing technology and the forthcoming one?" For UMTS operators this equates to "What can be done with UMTS that can not be achieved with 2G networks?" The answer is not very simple, as there are several other aspects beside the pure standardisation ones that impact on this issue. Differences between 2G and UMTS can be categorised in the following areas:

Differences caused by the technical distinction between 2G and UMTS standards

The main reason for the difference is not actually the bandwidth, which increases in UMTS to 348 kbps and later to 2 Mbit/s, but the cost of delivering that data and specifically the cost of delivering mobile data services as well as mobile terminal capabilities.

Differences caused by economic reasons

The differences here are caused by the cost efficiencies that will come from the volume of scale of delivering more bandwidth at a lower cost per bit. As UMTS becomes a mass market solution, some of the new technical developments that could be implemented both on 2G and UMTS will be provided only on UMTS because this is where the larger volumes will be.

Differences caused by other technical developments coinciding with the launch of the UMTS networks and services

A good example of this category is IPv6 (Internet Protocol version 6). This category is not so much about the difference between 2G and UMTS but of what will be available in UMTS networks that is not available in today's 2G and 2.5G networks.

Differences caused by business strategy

This category includes issues like new top line revenue models such as advertising. Also things like the evolution of the existing customer base or managing internal 'churn' into new services. It is obvious that new start-up 'Greenfield' UMTS operators have to have a new innovate value proposition in any developed mobile market.

Categories 1–3 will be examined in more detail in the following paragraphs. The fourth category (business strategy) is discussed later in the Competitiveness chapter in this book.

Differences caused by the technical distinction between 2G and UMTS standards

By the nature of its radio interface UMTS allows a higher bandwidth to be delivered to the mobile subscriber. However, many of these services do not necessary need UMTS speeds but can be delivered over GPRS already today. This is illustrated in Figure 9.1.

Because of the small size of the mobile phone screen, video streaming does not need high bandwidth. As an example, using 34 kbps streaming data, a 3 min 30 s video clip (music video) is good enough quality for say promotional purposes. The needed amount of data to be transferred is about 1 Mbit. If we assume that end user cost of the data will be half in 2.5G/GPRS on what it is presently in 2G Circuit Switched Data, the cost of getting this clip would be about 0.75–1.5 Euro[1]. In UMTS networks, the operator's technical costs will go down by up to 75% from the cost in 2G and will also enable the delivery of richer content to end users for the same price.

[1] 9.6 kbps * 60 s = 576 kb = > 3 min = about 1 Mbit assuming about 50% throughput rate from 14.4 kbps. = > 3 min = 0.5–£ CSD.

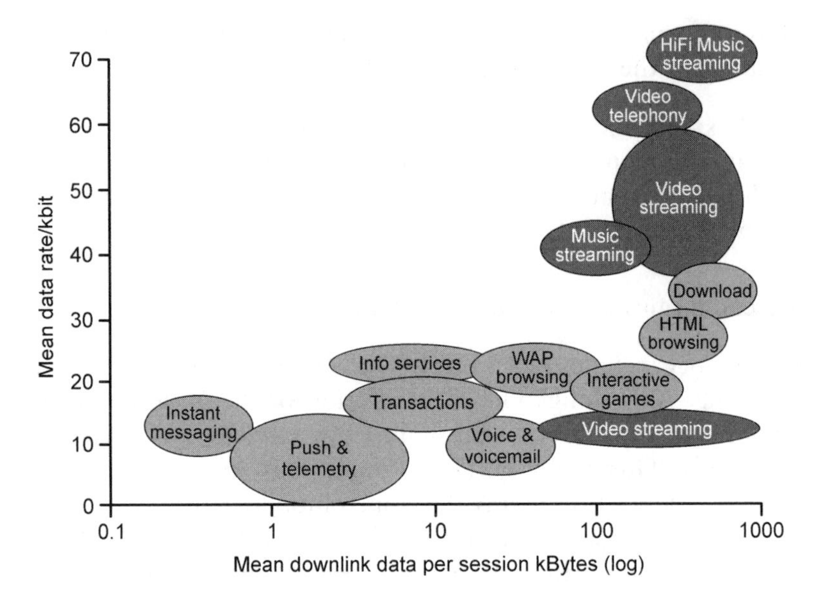

Figure 9.1 Bandwidth needs and date volumes for different services.
Source: Nokia, Make Money with 3G Services White Paper.

QoS – Quality of Service

There is also the question of Quality of Service (QoS). In 2.5G networks it is not possible to assign different QoS classes to individual services or packages. For this UMTS is needed. For example, in 2G a corporate client accessing the corporate Intranet would not get any higher priority to bandwidth than a schoolboy playing a mobile game even if the corporate customer was paying a higher access price. With UMTS, it is possible to assign QoS classes so the specific services or access packages take priority over the bandwidth resource so that higher service fees can be charged. This is one of the key differentiators of the UMTS technology.

Two of the biggest service beneficiaries in UMTS will initially be mobile commerce (mCommerce) and mobile entertainment (mEntertainment). Combined with new multimedia terminals the usage of things like mobile games and adult entertainment will boom and mobile phones will start to be used for the selection of goods and services, not just for paying for them. Reasonably priced video streaming will enable mass markets for mCommerce, sneak previews, WWW-radio listening and music downloads.

There are services that require higher bandwidth than can be provided by GPRS. Listening to real time CD (Compact Disk) quality music requires around 120 kbps and video conferencing requires around 100 kbps but these kind of services are not in the mainstream of mass market services. The level of usage of these kind of high bandwidth services is still unproven although the initial take up of the NTT (Nippon Telephone and Telegraph) DoCoMo UMTS FOMA (Freedom of Multimedia Access) service in Japan should give some early indication of what high end services will become mainstream.

The fact that high bit rate 2.5G requires complex implementation in the terminal side because of the utilisation of several TDMA (Time Division Multiple Access) timeslots is likely to result in a limit to the maximum data throughput in 2G networks. The initial implementation of GPRS and even cdma 1 × RTT are delivering bit rates between 20 and 40 kbps although this will be suitable for many of the initial mobile data services. When EDGE (Enhanced Data for GSM Evolution) networks are implemented there will be an increase in data rates per GSM (Global System for Mobile communications) timeslot but there will still be the need to aggregate timeslots. This means that services that require higher peak rates will become economical for the end users only in UMTS. This is a result of WCDMA (Wideband Code Divisional Multiple Access) and TDMA (Time Divisional Multiple Access) technical differences and is one of the key reasons for operators to move to UMTS.

Similarly the theoretical spectrum efficiency of WCDMA is about double that of 2G resulting in the cost of delivering data in fully loaded, capacity limited network is also lower.

Differences caused by economic reasons

There are some technical developments which will be implemented mainly to UMTS even though technically it would be possible to do so also for 2G. One important example is MMS (Multimedia MesSaging). MMS will be launched before UMTS networks are commercially available but because a growing number of the new terminals sold after UMTS launch will be UMTS enabled phones, it can be assumed that MMS will eventually evolve to be a mass-market service with UMTS.

The same logic can be applied to rich calls. Having available simultaneously, a speech call that can also support multimedia attributes like video or shared images will only become a reality in volumes with UMTS. Rich calls will enable the picture or video of the calling person to be seen, prompting menus for different alternatives for incoming call (answer, forward to message box, terminate, forward to special announcement), shared white board services and things like data search during the call. Other candidates that could become volume market implementations only in UMTS are Bluetooth, advanced location based technologies like E-OTD and SyncML for the synchronisation of different applications like calendars across a number of devices.

From an economics viewpoint, the R&D costs of bringing a new technology to market and the enhancements to that technology's performance runs into millions of dollars. So although we do not expect that current 2G technologies will cease to evolve there will eventually be a time that this evolution slows down and UMTS evolution overtakes the older technology. The cost of doing business will mean that in order to create the economies of scale needed to keep costs low enough and create mass market appeal the focus for development will shift to UMTS.

Differences caused by other technical developments coinciding with the launch of the UMTS networks and services

One of the most important items here is IPv6, the Internet Protocol that will replace IPv4 (Internet Protocol version 4). IPv6 was developed 20 years ago before the Internet was ever considered to be a mass market service. This new protocol provides operators with the possibility to offer QoS guarantees for end users and services. This is necessary to support video calls and real live applications over packet radio. This is also required for operator's ASP (Application Service Provider) activities since corporate users will request and be willing to pay for a specific guaranteed service if they run their corporate applications on an operator's mobile ASP.

9.2 *Why the Mobile Internet will be successful*

The industry has always underestimated the value that the end user places on mobility. Subscriber number estimates have constantly been too low. It has also been very difficult to guess which services will fly. For example many suppliers and operators did not believe in SMS. "You need to push one key 3 times to get one character, people will not do it...", "Why would users write text when they can just as easily call?" Yet nowadays the SMS business is worth billions of dollars globally. In Finland which has one of the highest service usage of SMS anywhere in the world there is no slowing down in sight for an ever increasing array of new services based on SMS. Typically SMS traffic accounts for about half of the average teenagers mobile phone spending in Finland. Even business people are using SMS more often since it is far easier and more polite to accept an SMS message and respond quickly especially if the called party is in a meeting. It is also faster and less expensive if all that is needed is the answer to a few simple questions. There are even managers who, due to the overload of daily e-mail traffic tell their organization to send an SMS if they want an urgent or in some cases, any response to their request.

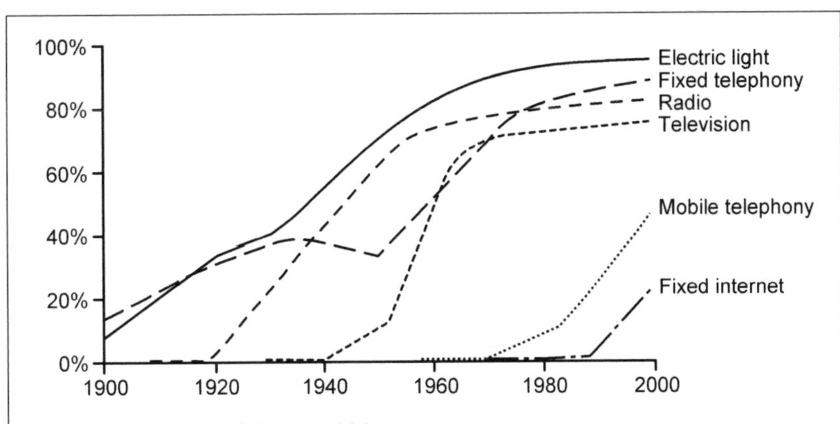

Source: US bureau of Census 1996
The diffusion of technologies into everyday use is happening more rapidly as technology becomes a main driver in society and consumers' capacity to adapt to their changing environment increases. The growth of mobile telephony has seen one of the fastest take up rates in modern times.

Mobile penetration is expected to be 60–70% in most developed countries when the first UMTS networks are launched. People today are much more mobile and Internet literate and are constantly requiring new services. These services will be developed not only by UMTS operators and suppliers themselves but by thousands of individuals and small companies around the world. The variety of services will explode. All this combined with technology development and operators' eagerness to make UMTS fly (to get a quick return on the license fees paid) should help guarantee the success of the Mobile Internet.

What are then the characteristics in the Mobile Internet that make people value it so much? There are 5 main categories, the 5 M's of Movement, Moment, Me, Money and Machines, as have been discussed earlier in this book. But apart from benefits to the individual users, there also are overall gains to business productivity.

Business productivity

One concrete reason for our belief in the success of the Mobile Internet is the productivity improvements in the enterprise business arena. Quicker approvals and notifications, faster responses to customer requests, faster problem solving at site are some examples where access to mission critical application are able to reduce the time that a normal business process takes. The Internet and corporate Intranets and Extranets have already been seen to improve corporate productivity dramatically. When mobility is then added into the equation there will be an added boost in a number of areas, most importantly in the access to up to date information, intelligence and the decision making process. For example if a factory manager notices that a component delivery is incomplete and wants to check what was ordered, he can currently walk to his office and access the order database and see what was ordered, before calling the supplier to complain.

Within the UMTS mobile world, the factory Intranet service would be extended to the UMTS terminal and the manager could access the information on the spot at the factory floor, speeding up the query and resolution. A sales representative can get instant updates on stock levels, customer history, market updates and even competitor movements, moments before meeting with a customer.

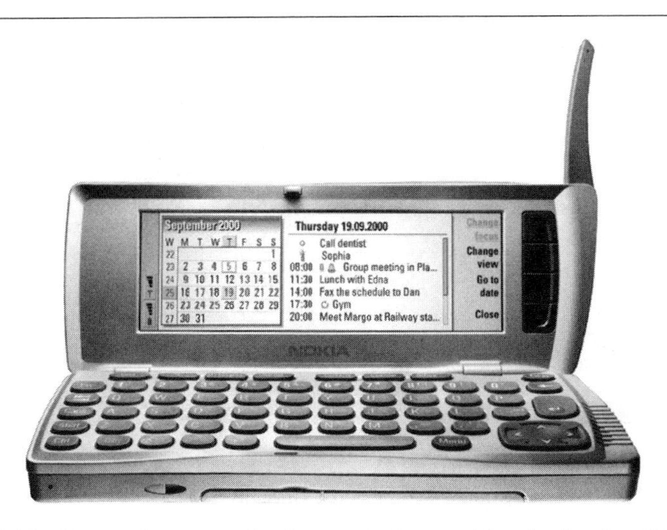

The Nokia Communicator was the first terminal to combine the functionality of a PDA with calendar and contacts directory and the mobility of GSM voice. As the communicator has improved in functionality making it an ideal business tool for many people, traditional PDA vendors are now bringing new product to market to catch by adding mobile voice and Internet capability to their product lines.

9.3 Evolution of services from 2G to UMTS

The transition to UMTS will not be a sudden leap from current technology and services to a whole new realm. Rather, the transition is best depicted by the term 'evolution', implying that UMTS will be built on existing technology. This has implications for today's end user services. The first mCommerce services have already emerged, built on 2G technology either using SMS or WAP. You can already buy a cola or a hamburger using your GSM phone or pay for a car wash without winding down the car window. The transition towards UMTS will spur intermediate services that will build on the characteristics and benefits of the technology involved. UMTS technology will finally enable the roll-out of a full mCommerce services portfolio, where the end user experience is hard to replicate in any other type of network environment. Ultimately, unlike in 2G and 2.5G services, UMTS operators and service providers will be able to offer services

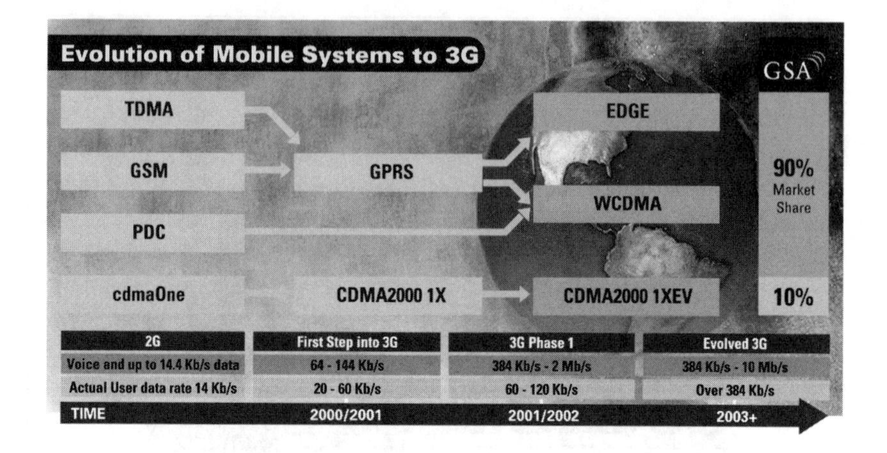

Figure 9.2 The transition to UMTS.

with little or no substitute and lock in customers through delivering an unprecedented service portfolio. This technology evolution is shown in Figure 9.2.

The discontinuity that will then be created by the transition to UMTS needs to be fully understood from both the impact on the technical ability of the network to deliver capacity, speed and cost efficient coverage as well as the service opportunity. This is the starting point for the development of a competitive UMTS service strategy.

9.4 Categorisation of Services for UMTS

A categorisation of services needs to be relevant to the entity which wants to categorise them. It is quite likely that a network operator's system may be different from the same operator's portal service's system, which again may be different from that of a content provider, etc. We have wanted to define one simple broad categorisation that covers all services, and provides some insight into the significant differences between those services.

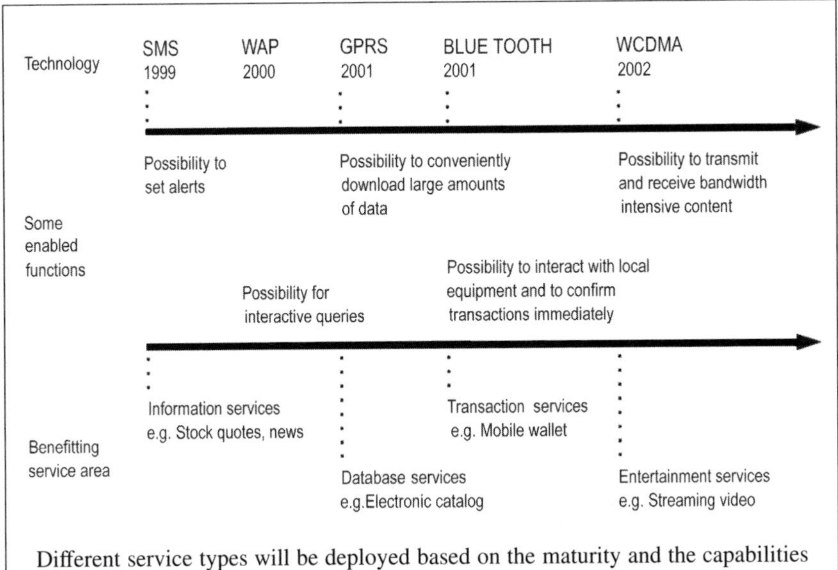

Different service types will be deployed based on the maturity and the capabilities of the different mobile technologies. Even some quite advanced services are possible with simple technologies like SMS. It is necessary to start early and also target the legacy terminal users who do not have UMTS phones or even GPRS phones.

Some of the major significant factors include who communicates with whom, what level of security may be needed, how is money handled, and how immediate do the services need to be. We have split the UMTS services into four groups: Inter-Personal Communication, Infotainment, Corporate Services, and Consumer Enterprise. Each of these has subdivisions, and of course this categorisation is but one of numerous possibilities. We will explore this division further in this chapter to illustrate the particular distinctions between each of the categories when comparing them against each other.

Most UMTS services will fall into four main categories as outlined in Figure 9.3.

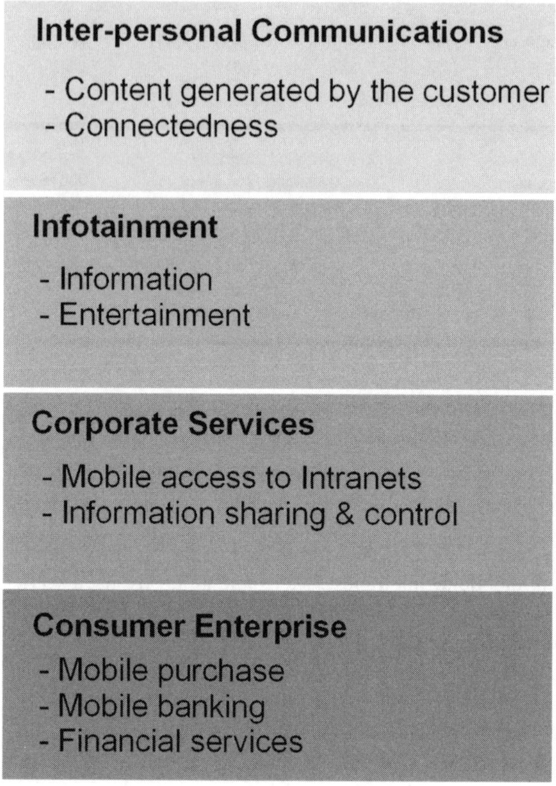

Inter-personal Communications

- Content generated by the customer
- Connectedness

Infotainment

- Information
- Entertainment

Corporate Services

- Mobile access to Intranets
- Information sharing & control

Consumer Enterprise

- Mobile purchase
- Mobile banking
- Financial services

Figure 9.3 Types of UMTS services.

9.5 Interpersonal communications

Interpersonal communications is about communicating with other people. The majority of this kind of content today is generated by the customer and is made up of voice calls and SMS. Almost all of the operator's business is coming from this kind of communication today with voice accounting for around 97% of all generated revenues globally[2].

[2] Source Nokia.

During 2002 there will be three market components that will change the face of interpersonal communications and create the change towards UMTS service adoption. Colour screens, MMS (Multimedia Messaging Service) and imaging phones.

MMS (Multimedia Messaging Service)

The biggest impact that we will see in the short term will come from MMS. SMS can be viewed as only the tip of the iceberg and the profit potential for operators. We say this because adding a picture to a mobile message is the next logical step. It is also simple and easily achievable. And it already happens in other media. We are accustomed to sending e-mail with attachments be it word documents, slide sets, excel files, images of our kids, the dog, jokes and animated cartoons. The behaviour already exists so it is a natural extension for this to happen in the mobile world.

Mostly MMS will be a 'fun' service where images are shared between friends and family. Mobile Internet jokes will be passed around and forwarded to others in the same way that e-mail jokes are shared today. Since much of this content will be customer originated there is little or no revenue that needs to be shared with other content providers. This means that operators are in an ideal position to create the mass market with attractive pricing.

Voice

We should not forget however that voice is still part of Interpersonal Communications and will remain so. Mobile voice has created the success of the market but it should not now be overlooked. There are still voice minutes in the fixed line business that can and will go mobile over the coming months and years. How much and when will depend on individual operator strategies in different countries but one thing is certain. To move more voice to the mobile phone business, operators have to compete with fixed voice tariffs that are often, but not always lower than mobile tariffs per say. This tariffing has to be implemented without reducing the 'value' of mobile calls when mobile.

To grow voice operators need to compete on price at specific locations where people prefer to use the fixed line. This means offering attractive pricing that competes with fixed line charges only in those places where fixed lines are used the most, the home and the office. By offering a 'Home' or 'Office' tariff solution operators can almost make the fixed line redundant for voice and remove the final barrier for all voice to go mobile. This is easier to achieve with consumers in the home environment than in the office where internal calls are free. However this should be seen only as a pricing strategy. Office zone to office zone calls could be free or at a single monthly fee per terminal where as all other calls are charged at the normal rate. Similarly any calls when placed from the mobile phone while at home; in the **Home zone**, would be priced at fixed line prices but all other calls are charged at the appropriate rate depending on the chosen tariffing plan.

Fixed access will then become mostly just for Internet usage at home when a large screen format is needed.

Text-to-voice

One of the primary solutions in multitasking is voice. If e-mails or news are read by voice applications and 'spoken' to the receiver, then e-mails can be received while doing other things, such as cooking, driving a car or gardening. Many forms of data might be able to be converted to voice translation and delivered in times when reading is not possible or even desirable.

Baby-sitter camera

A multi-session service could include access to the 'baby-sitter camera' which could be in the background while other services are handled so that one could always click onto the camera to view or if the baby cried the UMTS phone would automatically provide an alert. Leaving a 3 year old at nursery in a screaming fit (if you have never done it ask your wife to demonstrate the experience) can be less traumatic if from your UMTS terminal you can see that within 2 min of you leaving all was quite and calm. A door entry camera could be activated as soon as the doorbell rings and if you are not at home an instant connection could be made to your multimedia terminal so that you can deal with the visitor no matter where you are.

Location service(s)

Here we want to make a point. There is only one real 'location service'. It is called 'Where am I?' This service will pinpoint your location on a map, useful if you are lost or wake up in a strange place after a most enjoyable night out on the town. Most of the time however users want to know 'How do I get to…?' or 'Where is the nearest…?' So when you hear the term location based services (plural) then we believe this means 'services that use location as one part of the interaction.' Not a service that is just about location.

The type of service that fall into this Interpersonal Communications category then are those services that provide the freedom of movement while still allowing us to remain in contact with other people.

9.6 *Infotainment*

Any service that entertains or provides information falls into the category of Infotainment. Games will be one of the major services in this area as will be music, community services, jokes, news, sports, fashion and even adult entertainment. In fact any content that informs or entertains can be found in a newspaper or magazine can be included in Infotainment. A mobile What's On? service would be typical for this category. The UMTS operator can deliver information about what is on the TV tonight, what movies are showing this week at various cinemas, what is on at the theatre and are there tickets available? What bands are due to perform in what towns or countries over the coming months? Clearly services in the Consumer Enterprise category, especially payment services can link very well with Infotainment services if tickets need to be purchased or a new gaming download is needed to continue to the next level of the mobile game.

Music

Music has the potential to be one of the largest revenue components in Infotainment simply because music is already portable and is increasingly becoming integrated into mobile phones. Delivering music in digital content to a mobile device that becomes part of a secure, pirate proof distribution channel will remove a large part of the music distribution chain and along with it costs. Current music stores will not disappear overnight but we do expect to see a gradual shift towards the digital distribution of music and the youth market will be the early adopters. This group are also the ones who spend a higher proportion of their disposable income on music than other older segments of society.

Music fan club

A typical subscription promotion service would be a music fan club. A small subscription of say $1–3 would result in a regular news message about the band, details of when and where the band is playing with preferential access to seat tickets and instant download of the latest music video. Many users would accept a limited amount of relevant promotions if they received credits or discount coupons. Free SMS news is already happening in Europe if the user agrees to one SMS advert per day so there is evidence that mobile advertising will happen.

Jokes service

If Tomi wanted to sell his jokes over the Internet how could he do it?. He could just set up an Internet site and start selling his best jokes for a couple of pennies to each visitor, but the cost of administrating the billing would be too restrictive. Instead he must either sell monthly subscriptions packages or sell advertising space based on click-through rates. Yet click-through rates in the Internet, where users click on an advertising banner to get more information, are 1% and falling. In fact recent evidence appears to support the feeling that regular Internet users are less likely to click through on a banner than new converts, so click-through is not a realistic way of calculating the value of advertising on the Internet today.

13 Vignettes from a 3G Future

m-Cheering, m-Booing

I really like the way we can get immediate feedback now with the UMTS voting services. We use it all the time in our events and conferences, having the audience give immediate feedback with their UMTS phones. The voting is free and we have a prize draw at the end. We project the results onto a big screen and the speakers call it m-booing if they are not popular, and m-cheering if things went well. We find that we get wonderful real-time feedback and audiences are much more eager to provide the feedback with this medium rather than the old-fashioned paper performance evaluation.

Various forms of m-voting and m-polling services will become commonplace after a large enough penetration of mobile Internet devices. As various companies, services and organisations strive to become more responsive to their customers, interest groups and audiences, the immediate, digital form of opinion collecting through mobile terminals will become more and more common.

The only option for Tomi is to sell a periodic subscription at a one off cost that can result in some reasonable profit, but since this reduces his target audience to those who will commit to say a year's subscription it is unlikely to make Tomi an e-Millionaire. This is where aggregated micro-payments in UMTS come to the rescue. A joke service where a user gets today's joke sent direct to the UMTS device every morning, no matter where he is could be charged at say $0.5 per month. This appears so low in cost that many people could be interested in it. If the user could select the joke segments that are of interest, the value of the service could increase to say $1 per month.

Bandai services in Japan

From a business perspective we doubt that any bank manager would get excited about one dollar per month per subscriber but lets consider the Bandai service provided by DoCoMo over i-mode in Japan. There are over 2 million Bandai subscribers who pay 100 Yen, about $1 per month to download every day a new Bandai screen character. This sounds like a totally unlikely UMTS service but the revenue totals 200 million Yen per month, which is $20 million per year. This service is provided with a staff of about 10 people, and returns a very nice profit.

9.7 Corporate services

The most obvious Corporate Services are e-mail and access to Intranets and Extranets. These offer the basic extension of the office into the mobile world. However there are many other services that fall into this category that will find a growing amount of corporate usage.

The most basic of these is access to the **corporate phone book**. Being able to access any number for any employee in the company is one Corporate Service that is possible now. With UMTS and MMS it will be possible to include a picture as part of the digital business card making the transaction just that little more personal.

CRM (Customer Relationship Management) is growing in usage and will become a major tool for door to door sales people to top account executives, all of whom need to know the very latest situation about their customers or potential customers. CRM was discussed in detail in the Movement Chapter.

In the Corporate Service category security is the one feature that has to be supported and guaranteed. For all Corporate companies security is the one area they can not compromise on since malicious attacks against corporate systems happen not only daily but hourly all over the world. For this reason, many companies are only looking for 'bit pipe' access or service from their mobile operator. But this level of service should not be underestimated for its potential to create a strong and reliable revenue stream. The confidence or lack of it, that many corporate customers have in mobile operator's ability to deliver high quality datacom services can be changed over time. Starting out as bit pipe supplier with a few simple corporate services is the way to develop a better understanding of what fuels these corporate fears and eventually over coming them.

Micro-payments for corporate customers

Corporate users who travel, we believe, will quickly see the potential of micro-payments to make their life easier. After even a short trip the time taken to fill out the **expense claim** is frustrating. If your secretary does it then it is just frustrating for someone else. The mobile operator could provide a billing solution where the traveller could define a cost code for each micro-payment and these would then appear on the mobile bill as separate items under a travel heading. The more automatic the purchase code allocation the better. No more travel or expense claims to fill in. Yes please.

9.8 Consumer enterprise

We have little doubt in our minds that at some point in the future all financial transactions will be performed using a UMTS terminal. That day is probably sooner than many imagine. Consumer Enterprise services are those that enable any financial transaction to be completed. Many of these services will have a direct link to other services providing an enhanced user experience.

What's On?

So a What's On? service may provide information about movies at the local cinema and the payment feature of the Mobile Banking service will enable the UMTS consumer to purchase tickets and have the money deducted from their bank account. This may not always be the case. The cost of the tickets may be simply added to the mobile phone bill. However, this will increase the operator's exposure to debt and with some customers this exposure could come with a high risk of default. In these cases it is better to take a debit card approach and deduct the money directly from the customers bank account.

Direct bank payment

Direct bank payment via the UMTS terminal is also one way an independent portal can collect money for goods or services it offers. There will be many content providers that do not have a billing agreement with the UMTS operator and they will require an efficient way to collect money at the time of sale. This is where extended financial services will help grow the overall business for UMTS services and make mobile electronic purchasing commonplace.

Charging information as a UMTS service

For the network operator, the micro-payments set-up does not necessarily represent a strict choice between producing or not producing content. The operator itself may very well produce the content in the set-up. However, there is already evidence that even the micro-payments set-up as such may provide an operator with a feasible business case. In Japan, the i-mode service of NTT DoCoMo is based on aggregating charging information. Customers are able to access thousands of sites produced by numerous content producers. i-mode services used will be charged on the customers' phone bills alongside with other services customers may use (e.g. normal voice calls, provided by NTT DoCoMo).

As always there will be people who will question these changes and often use reasonable arguments to justify their case. The problem with these arguments is that the rational is based on the past and here we are talking about the future. MP3 is a typical example. Digital music delivery will happen, there is no doubt about it, but the music industry will fight the change until they can find a suitable business model that allows them to make money and retain some form of control. The same can be said of many other industries yet it will be the visionaries who predict and ride the tide of force for change within society that will be the winners. These services don't always have to be sophisticated.

Micro-payment purchases

We can identify different types of micro-payments. The end user purchases a service, lets say a **yellow pages search** from the operator and the operator bills $0.10 for it. The user sees that this amount is added to his regular monthly telecommunication bill. From the system point of view the billing doesn't differ from today. The billing logic is the same as in current services.

Micro-payments become more challenging when someone other than the operator is receiving the payment and especially if the user is roaming onto another UMTS network. The operator can act here as a payment intermediary and take same kind of commission similar to how credit card companies make money on transactions. Operators can aggregate the service from all the users and transfer the payment to the service provider say every month, less their commission. In a roaming situation the operator has to have an agreement with the roaming operator as to the percentage commission paid and has to decide if there is an additional roaming user cost or if there is double commission deducted from the service provider's payment. Users today pay a premium for roaming voice so it is likely that they will accept a higher cost for roaming data services in UMTS.

Pre-paid users which make up an increasing percentage of an operator's subscriber base pose some difficulties for micro-payments. To accommodate the cost of higher transactions with mCommerce, pre-paid subscribers would have to purchase more or higher value pre-paid cards. Alternatively we will see the proliferation of mobile payment solutions like the one that the Philippine's mobile operators SMART has deployed. In this case the subscriber can load the mobile account with money directly from the bank account using a mobile phone and SMART Money account. As long as the SMART Money account is in credit the subscriber can pay for goods and services, even those that are of small value.

Family finances

Lets extend this to the family. The UMTS operator provides a service for family accounting. The monthly mobile bill then becomes part of the family accounting solution where it is relatively simple to see where the money is spent and the spending patterns of each family member. We can give our children a spending allowance that they use via the UMTS terminal and have better control and visibility of where they spend their money and on what. By taking cash out of the pocket and putting it into the terminal we get back some level of control.

mWallet

We may not see in the near future the total cashless society but UMTS micro-payments and mWallet type of applications where the subscriber can download cash to a secure SIM (Subscriber Identity Module)-card and pay at shops using Bluetooth, will take us some way towards this solution. Although there are already pilots of electronic cash it may take a long time until one de-facto standard emerges and critical mass of shopkeepers have systems to accept it. Standardization here has a big role to play.

The operators' competitive advantage in micro-payments rises from their relatively cheap billing costs. For larger payments the credit card companies and banks may still hold an advantage because of their currently better capabilities to control credit risks. It is likely the users will hold their credit card details as part of their user profile and pre-define what level of individual spend will go to the credit card and what will be added to the mobile bill. Obviously micro-payments of a few dollars will always go to the mobile bill but the user may define that any payment over say $40 would always go to the credit card.

Since we believe it is likely that a large number of customers are willing to pay something for even relatively small pieces of information these smaller payments will amount to profitable services with large audience usage. Consider today's cash usage patterns. Items costing under $14 account for some 22% of cash spent globally[3]. (the UK credit card companies' definition of micro-payments is approximately $17). This audience and revenue stream can be reached with a UMTS micro-payments solution. Both content providers and network operators will leverage their existing assets and end-users will get services they require. An important fact is that any micro-payments solution has to have a win-win-win potential. Win for the service provider, operator and customer.

As the level of revenue substitution increases and more and more payments accumulate to the mobile bill we could see some operators partly becoming banks. This will not happen overnight. It takes 5–7 years to get a change in behaviour patterns but if you ask people if they would like just one bill for everything and one payment each month, you will find that a single monthly bill is the preferred option for many. This can be seen today with the finance companies marketing the consolidation of all debts into a single once a month payment that is less than current cumulative payments.

mSalary

Lets end this section with one radical and far out idea. We like far out ideas since it stretches our mind. Your salary is paid into your UMTS account every month. You pay for everything via your mobile phone. All your monthly debts like heating and lighting are sent direct to your UMTS account electronically. You are notified the instant they arrive. You authorise and post date the payments from your UMTS terminal whenever, wherever you are. You don't get letters anymore you get electronic postcards. Promotions are delivered direct to your UMTS terminal and are relevant to your lifestyle. In this radical world, where is the postman?

[3] Business 2.0, Future Publishing, Bath, October 2000.

14 Vignettes from a 3G Future

Downloading the Newest Music CD

The rock concert is about to start. I was surprised to hear the announcement that the band has arranged that their new CD can be downloaded now while the concert is going on, for half price, only for those who are attending. I love this band and don't have their new CD yet. It is much cheaper to download the CD now onto my phone than it would be to buy the CD at the store. And its much more convenient than downloading the CD to my PC and then transferring the songs to my phone. I use the phone as my music player already anyway.

The live sending of a song or whole CD - in CD sound quality would require a high quality class and use a lot of bandwidth. But sending the songs in lower quality class and over a period of an hour rather than some minutes, and allowing for delays, and taking advantage of simulcasting, would make mass deliveries of music CD's to concert audiences a very low cost delivery option. The record company saves in printing, packaging and sales costs. Part of those savings could be passed onto the fans in attendance. The fan gets also a direct contact from the artist. And the artist gets a database of direct UMTS phone numbers of its most eager fans, those attending concerts and buying its music. This database would form the core of its future marketing efforts, such as "since you are one of our most valuable fans, we want to send you the ring tone version of our new hit a week before the single hits the record stores" etc.

Categorisation

The categorisation we have suggested is not the only one, and by no means the one best for all situations. For the purposes of this book it is as good as any other. We have tried to illustrate this basic level of categorisation by what they include and what they do not. And we hope we have shown by means of examples, some real and potential services that can fit this way of categorisation.

9.9 Psychology of service creation

Ask yourself these questions:

How many engineers do we employ?
How many Marketing people do we employ?
How many psychologists do we employ?
How many sociologists do we employ?

It is likely that the answer to the first two questions will be easy to give. But we are prepared to bet that most companies, be it operators, content providers, developers or others have in the main, no psychologists or sociologists or anyone in their organisation who fully appreciates behaviour and society changes. There is no one feeding this vital intelligence into the company. No one helping the organisation fully appreciate how customers behave when they use their mobile phone, what type of behaviour they have at different times of the day and in different situations with various people. Yet, as we move progressively into the world of the Mobile Internet it is becoming imperative to appreciate what turns your customers on! Knowing what they are attracted to, what their impulses are, how they feel, what makes them happy, what makes them sad, what makes them take action and what makes them procrastinate. Knowing all these things can increase an operator's competitiveness. When customer attitudes are fully understood it is possible to create the most desirable mobile data services and then tailor them to meet customers desires, needs, moods, cravings, whims, and fantasies.

In the past the mobile phone business has been about longer battery life, smaller phones, weight and to some extent design. Now it is about neater and more funky phones, a better user interface, larger colour screens, better messaging capabilities, customised ring tones, gaming, images, e-mail and brand. With UMTS, knowledge about services, content and customer needs and behaviour become the key competitive advantages for operators.

Lets take a real life example from the food industry. One large supermarket is reputed to have studied the buying habits of its customers and found that fathers would come into the store late in the evening and just purchase a pack of diapers. The analysis was that these customers were probably shopping on the way home from work with a request from their wife to buy diapers since they were running low at home. One smart employee then decided to place a shelf of beer next to the diapers. These fathers then on many occasions purchased diapers and a 6 pack of beer. The key here is that as a marketeer the knowledge needed in this situation was the psychology behind changing a diaper from the male perspective. Changing a very smelly diaper after a hard days work can sometimes only be compensated by drinking a cold beer. The secret is to create action that leads to compulsive consumption.

In a similar way you can now find Coke in its traditional place next to all the other soft drinks but also located next to spirits where is can be seen as a mixer and next to the picnic and party food where it is positioned as one of the drinks for these occasions. Coke is understanding that their drink can be used in many different situations and locate their product in places where it can be associated with these occasions.

It is the same with UMTS services. By understanding the occasions and situations that customers are in at any one time and place operators can more accurately define what services are most appropriate and more importantly what supplementary services may also be sold based on what their customer actually does.

To browse or not to browse?

Everyone agrees that the success of UMTS will come from the rapid adoption of new data services. That however is not the full picture. The success of UMTS will come with these new services and their growing usage but also their usage numerous times during the day.

One of the things that we can learn from WAP is that customers will not browse from their mobile phone 5 or 10 times per day, every day. The i-mode experience which is just that, browsing on a regular basis is not expected to be repeated globally since there are specific cultural differences in Japan like limited penetration of home PCs and the fact that many people commute up to 3 h every morning and evening giving ample time to search for new and interesting content via the mobile phone.

Most of us will need to be taken through the buying cycle before we will consume any new service. One view of this buying cycle is referred to as A.I.D.A. Awareness, Interest, Desire, and Action. Lets assume a man sees an attractive girl in the street. If he walks up to her and kisses her he will at the very least get a slap to his face and maybe even arrested. A better approach would be to create awareness in the fact that he wanted to kiss her, create interest in his kiss, then desire that this kiss is something to be wanted and finally action that this kiss is acceptable.

It is not enough to think that customers will browse for content just because it is there. We would be waiting a long time if we just sat on a park bench and waited for the girls to come and ask us for a kiss. In the UMTS services environment it is essential to create first the awareness of the content offering and then build the interest. This could be a simple push news headlines service where Awareness of the news headlines is provided, Interest in the story is created and Desire for more detail is instilled in the customer creating the Action to download the full story. In a similar way if a customer orders and pays for a new ring tone of the latest song from say Madonna a music promotional service could let this customer know that the CD is available (Awareness) at a discount (Interest) can be ordered now and delivered electronically to the terminal or hard copy to any address (Desire) and it can be paid for by pressing one key. This is likely to lead to Action to purchase which is what the operator is trying to achieve.

Current mobile phone customers use their mobile phones primarily for voice communications. It is unrealistic to expect them to just take up and actively consume new mobile content just because it is there. It will have to be pushed to them. There is a great deal of what is known in the marketing world as 'noise'. This is other advertising or in our case also other existing sales and information channels that pull the customer's attention in various directions diluting the buying process. Fortunately the mobile phone is becoming a primary delivery channel for more than just voice services. It is also the one channel that can be personalised and customers are more likely to trust it and the content delivered over it than other media. So if used in the right way it will be possible to move customers towards using their mobile terminal in new ways.

The youth have it

The older generation, and by that we mean anyone over 25, have been through at least one major change management in their lives. Many of us have been through three or four. The transition first from the manual typewriter to the electronic typewriter, then to the first desk top computers, then to local networks and e-mail, the Internet and Intranets and now the mobile phone and Mobile Internet. What we find is that the older we get the longer it takes to change our behaviour. We also find it difficult to appreciate what drives the younger generation. Why do they listen to rap music or music with the same continuous beat and rhythm? But then our parents said the same thing about rock and roll and heavy metal. Times don't change quite as much as we think.

Clearly understanding the youth market will be key to creating a competitive offering in UMTS. In fact we would even suggest testing many of the proposed UMTS services that are being developed on your teenage kids at home.

Appreciation of the youth market and what motivates them is fundamental to the long term success in UMTS. Imagine what it is like for a teenager today with a mobile phone. This social group can be more or less 100% penetrated. The mobile phone is frequently their main communications device with their family and friends and SMS can often be over 50% of their monthly bill. Teenagers are driven by style, fun, uniqueness, fashion, image and a desire to be individual. They don't just want to be one person in a crowd.

As these people age they go on to college or university and take their mobile phone with them. It enables them to keep in close contact with their peers and family while they are away from home and it helps them to be part of the new community that they have just entered. When this young adult graduates and enters the work place they are bringing with them new ways to communicate and do business. In other words, it is less likely the organisation can dictate all the rules to them. This group of employees often have more advanced rules that are more efficient.

Their first pay cheque when it arrives is not spent on a broadband access to the home and a new top of the range PC. Instead they go out and celebrate. Their UMTS mobile phone is their primary access to content and many of this society group will never purchase a fixed line for voice. The opportunity then for the UMTS operator is to learn their language, get under their skin and build services that address their early lifestyle needs. As their lifestyle evolve the services that are offered also have to evolve so that compulsive consumption is part of their everyday life.

9.10 Typing up types

The reason we categorise services is to help deal with large groups of similar services. Much more important than one absolute categorisation, is to find a suitable one for our needs. First we must understand that while for the end-user most UMTS services will seem similar to those that can be deployed on current 2G and 2.5G networks, there are significant differences and in particular the cost efficiency of UMTS networks will prove an overpowering cost-benefit to UMTS operators. The difference is not unlike comparing passenger jet airliners of the first generation, such as the Boeing 707 and McDonnell Douglas DC 8 and the French Caravelle to the second generation wide body jets, such as the Boeing 747, the DC 10 and the Airbus. For sheer cost-of-delivered-passenger, the jumbo jets win out on every route where there is enough passenger traffic to fill the planes. Similarly on all services where demand will be large enough, UMTS networks will provide overwhelming cost-efficiency benefits to second generation networks.

In addition to understanding the differences to older technologies, it is important to be able to categorise the actual UMTS services. The UMTS service universe will be vast and each player will find ample opportunity. The key is not to wait too long, and to try many opportunities. Not all service ideas will be winners, but by trying many, there will be several successes. To quote the first century Roman poet Ovid who said: "Opportunity has power everywhere; always let your hook be hanging – where you least expect it, there will swim a fish".

10

Nothing is more practical than good theory.'
Joseph Grunenwald

Marketing UMTS Services:

Segment, Segment, Segment!!

Timo Rastas, Jouko Ahvenainen, Michael D Smith, Tomi T Ahonen and *Joe Barrett*

Referring to the introduction chapter, we need to use current marketing theories to achieve the best success with UMTS (Universal Mobile Telecommunications System) services. A UMTS operator will be facing fierce competition, from more competitors than it is use to, with an untested technology, in an environment that allows for totally new services. The UMTS operator will need to recoup heavy investments in the infrastructure and in many countries also to recoup high licence fees. Operators may also have owners who wish to achieve a fast return to their investment. All of these factors place higher value to effective marketing of UMTS services. To understand the marketing environment, we start by examining the elements which are changing drastically, and then build a framework from which to provide perspective on the change by building an analogy from the retail industry.

10.1 How the marketing environment evolves with UMTS

The mobile service provider's business is changing dramatically with the introduction of mobile applications. The service providers offering has evolved from basically one single service, 'the connection' between A and B subscribers, with different tariffing schemes, the 'products'. Service providers now offer a continuously increasing number of distinct services and in UMTS this increases exponentially. In fact so many new services are being created that calling them mere VAS (Value Added Services), implying that the services only add value to the core voice call service, is not justified anymore. What used to be VAS and generated less than 5% of the average monthly revenue per user, has grown in some segments to a sizeable chunk of the monthly spend. In addition, the sheer number of available services is growing at a dramatic pace. The increase in services is driven by three discontinuities: advances in terminal technology, tariffing structures based on value or transferred data rather than connection time and access to the vast pool of digital content and applications available on the Internet.

The mobile terminal is rapidly adding capabilities. Advances in display technology have brought colour screens with increased graphic resolution to the market that enable new content forms. Energy efficient processors have opened up mobile terminals and PDAs (Personal Digital Assistants) for programmers. New terminal capabilities, such as video capture or music players and interfaces, such as Bluetooth, unbind the input/output restrictions previously imposed by the user's thumbs in keying and text displays.

Packet networks change the economics of usage. When users are no longer penalised for every second they spend online the experience will improve dramatically. Access to the Internet's vast pool of digital content accelerates the speed at which new services will be introduced. Even though few Internet applications or content can be directly used on mobile terminals in their current form or with the same convenience as on PC (Personal Computer)-based browsers, reformatting and redesigning the service concept to fit mobile use will change the way that people access and utilise content in the future. Accessing new digital mobile content and especially the use of permission based push delivery will drive the take up of UMTS services. More and more content providers and application developers are starting to realise what is needed to drive acceptance and take up of new services and are working towards early introduction of their ideas for new content or new service delivery solutions.

All these changes rapidly increase the number of service available on the market through mobile operators, through Internet service providers or through businesses. At the same time the typical life span for mobile services will be dramatically reduced. Services that rely on user supplied content are inherently more long term than those using content provider content. Withdrawing a service into which users have entered ongoing information over a long period of time is a long term project which requires an alternative service to be adopted. On the other hand, withdrawing a service with no user content, such as many information browsing services is rather straightforward.

10.2 Retailer analogy

Within the UMTS services marketplace each UMTS operator will need to manage thousands of services. That may not seem odd until we recognise that current mobile operators typically manage only a couple of dozen services. The change from the current cellular operator to a UMTS operator is one which could be compared to the consumer retail industry. The analogy would be similar to a kiosk manager becoming a mega-market manager. The kiosk manager would need to manage a handful of different products such as cigarettes, magazines and candy. The mega-market manager would need to manage dozens of different types of orange juices, several brands and types of milk, etc., and products would be categorised by types. To make the analogy more relevant we should talk about a 'mega-market' – a very large supermarket.

The similarities of managing a Mega-market and managing a Mobile Portal are quite significant helping to make the analogy strong and useful. In the following table we can compare similarities of an operator's portal business and mega-market (Table 1).

Table 1 Comparison of a mega-market and mobile portal

Mega-market constraints	Mobile portal constraints
Patronage of store and its competitors	Churn
Location	Screen size
Size of store	Bit rate of data
Appearance of store	Ergonomics of terminal, software
Lay-out of store	Personalisation of portal
Cleanliness	Security
Speed of service	User interface
Parking	Addresses
Payment methods	M-payment

Retailing as a model for services retailing

The mega-market retailer who is managing thousands of different products, called SKUs (Stock Keeping Units), experiences failure rates in excess of 95% with new products, and has a shelf space that is limited and needs to be continuously optimised. There are a striking number of similarities between this business and the UMTS operator business. The UMTS Mega-market is a marketplace for UMTS end-user services. The operator's main task is to create a platform where different service providers can bring their mobile enabled content and services. The operator creates the portal based on their own brand but uses other brands to highlight the quality of the marketplace. The operator can offer the service providers' visibility to their customer space, billing and customer services. On top of that they can provide the end-user's location or preference information to service providers when required so that developers can delivery the content that is both relevant and needed. Figure 10.1 illustrates the operator's 3G Mega-market portal plan where some of the similarities with the mega-market business can be seen.

Mega-market own brands

In the same way as markets offer their own branded goods the operator can offer services with their own brand. All communication related services for instance could be brought under the operator's own brand. Other sticky services such as calendar and storage space

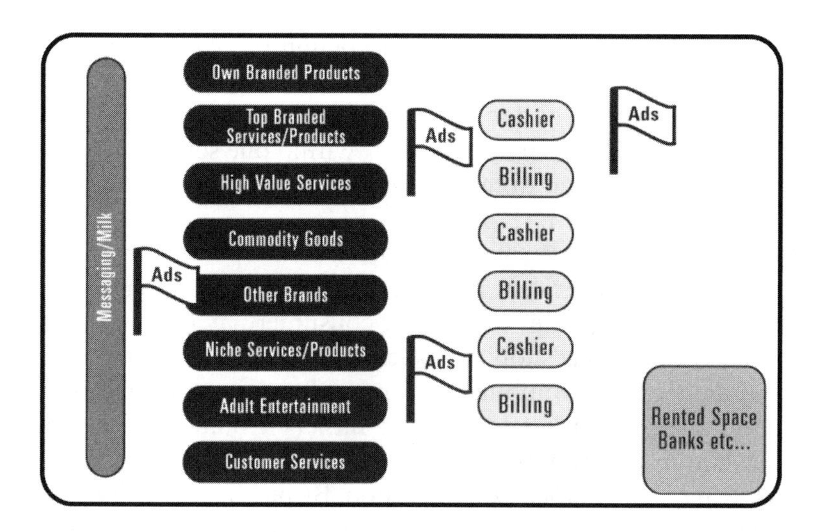

Figure 10.1 3G Mega-market Portal Plan.

for personal data can be provided by the operator to minimise churn. Customers are less likely to change operator if they need to leave their contact information and personal data behind and start building up a new user profile for instance. The UMTS operator can learn a great deal from modern retailing management practices such as loyalty schemes and promotional campaigns. For the UMTS services retailer the need is to focus on managing categories, managing user segments and managing the store (portal).

Category management of wireless services

Category management recognises that we cannot effectively manage the service offering on the level of services nor can we effectively manage at the portal ('the store') level. A category is a distinct, manageable group of services that the users perceive to be relevant or substitutable in meeting a need. Hence, a category is defined by answering the following questions:

1 What is the need users want to fulfil?
2 What services provide a similar solution to that need?
3 What services does the user see as relevant of substitutable in meeting that need?

Note that there is no single standard definition of categories. Every service provider can define the categories to best match the needs of their target customers and to reflect their service creation and delivery strategy. The categories may even vary within one service provider as the needs vary between different user segments, e.g. youth and business users.

Mega-markets have a shop structure which separates the products in the aisles, e.g. beverages, pet care, household etc. often with signs above the shopping aisles to help the consumer navigate in the shop. But it also helps the shop management to sell more products by placing similar or complimentary product next to each other. In one case a mega-market found that fathers would visit the mega-market late in the evening and would only purchase baby consumables like nappies (diapers) or baby food. By placing the baby product category next to the alcohol category this consumer group were more likely to also purchase alcohol when on this specific visit.

Mega-markets have been an early adopter of technology to enable them to identify spending habits and know the money consumers normally will allocate to different categories (not necessarily to a specific product as such). The loyalty card also helps the mega-market build up a specific profile of the shopping habits of each consumer and how they spend their money in the store. In a similar fashion, UMTS operators need to structure mobile services into a hierarchy of categories, sub-categories and segments – not on the basis of technology but user needs and habits. And operators already collect usage data by generating so-called 'CDRs' (Call Detail Records) and other data into their billing system.

It is useful to illustrate the definition of a category by analysing the user need to send a message to someone else. There are several similar technical solutions that could fulfil that need: chat, SMS (Short Message Service), e-mail, fax, voice mail and the user need may not match exactly the technical options. The user's secondary decision making criteria may be based on the type of message and only then a choice between the available service will be made. This is demonstrated in Figure 10.2.

Since categories bind user needs with services they also link the spend that users' allocate for fulfilling those needs with the potential revenue that can be earned from services. It is also important to remember that from the user's view a category may include services

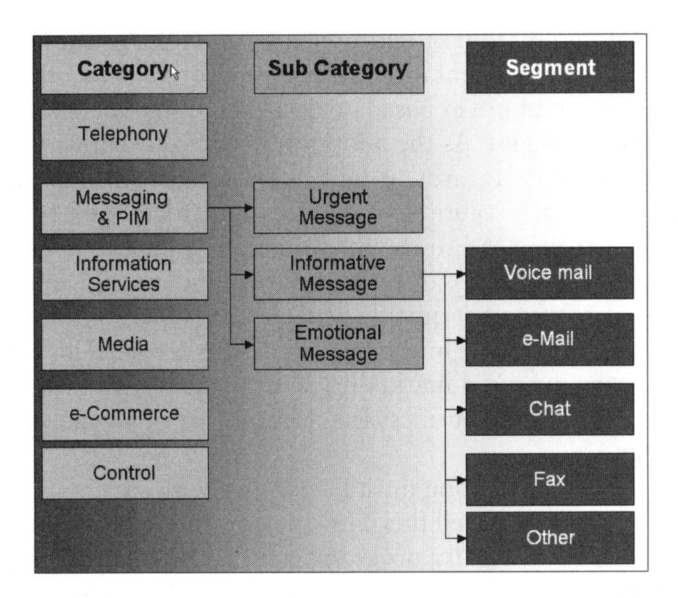

Figure 10.2 Need fulfilment of a message sending decision.

or products that are not wireless services like with music which is a category where mobile services are only starting to participate in the available revenue pool.

Customer relations

Customer relationship management is traditionally an asset for the operator which has already established itself as the customers' trusted billing partner. This advantage will again attract content providers and fill the store with services, so there will be a tremendous opportunity to collect customer-related information. Real world retailers compete with their customer friendly attitudes and their ability to listen to the customers needs and complaints. There are many cases where a complaint like the return of faulty goods has been handled badly by the retailer and results in the customer never returning to that store. The UMTS operator's customer relations mindset will have to reflect that of successful retailers in the market today if they are to compete successfully in a competitive market where the other operators are eager to eat your lunch.

Natural accumulation of end-user data provides greater possibilities for customer segmentation and an opportunity to offer service varieties at different prices based on end-user requirements e.g. classic vs. premium packages. As the adoption of services will vary between early adopters and laggards, it will be of paramount importance to maintain constant evolution of services and focus on crossing the chasms so that the road to the mass-market population will be well paved. Some operators will find that their initial segmentations or categories were wrong or quickly become outdated. For these reasons it is necessary to evaluate how effective these areas are. This is nothing new since carrying out a marketing audit on a regular basis to assess the effectiveness of the business is a natural part of any good marketing strategy.

All customer information must be handled with care, as no one can really own, let alone re-sell this data apart from the customer himself. Nevertheless, using loyalty programs and various directed discounts and promotions is a huge opportunity, when the customers permit this on their part. However, the lucrative mobile advertising market should not be spoiled by an uninvited flood of promotions. A more difficult area, the handling of complaints and guaranteeing the quality of service have to be managed efficiently. The Operator is the owner of the store and hence perceived responsible for all its products and services, regardless of how much actual control they may have over their functionality especially ones provided by third parties.

Applying the rules of the retailing world

The business practices in the retail world are in many aspects transferable to the emerging UMTS marketplace for mobile portals. Competition with other portals and independent service providers will be tough, so operators will need to excel in the eyes of their customers armed with much of the same weapons as today's Wal-Marts, Benetton's and McDonalds. In addition to attractive prices and superior quality, the mobile mega-markets will need to demonstrate real added value in their product portfolio. The usually free service portals in the fixed Internet will be acting as competing 'discount stores' and they will constantly challenge the operators with their established brands. However, operators have exclusive competitive edges like micro-billing, mobility, location-relevancy and personalisation.

Operators will be able to benefit from their existing customer base, which will easily find their way to the operator's portal store, at least as a first stop. To keep them interested and to invite frequent visits, the operators must offer a wide variety of exciting and evolutionary services including brands as well as fads, which are fashionable novelty services with a more shorter life time. A successful mobile operator's mega-market portal will generally offer such prominent visibility to consumers, that there is likely to be enormous demand for the shelf space from service and content providers. In allocating the portal space operators will hold a substantial amount of power and hence their position in revenue sharing, joint marketing and customer care negotiations will be strong. Examples from the retail world reveal that mega-markets do squeeze the brand margins, they do require joint marketing and they do quickly eliminate unsuccessful propositions out of their product portfolio. Figure 10.3 illustrates the breadth and width of the potential portal content.

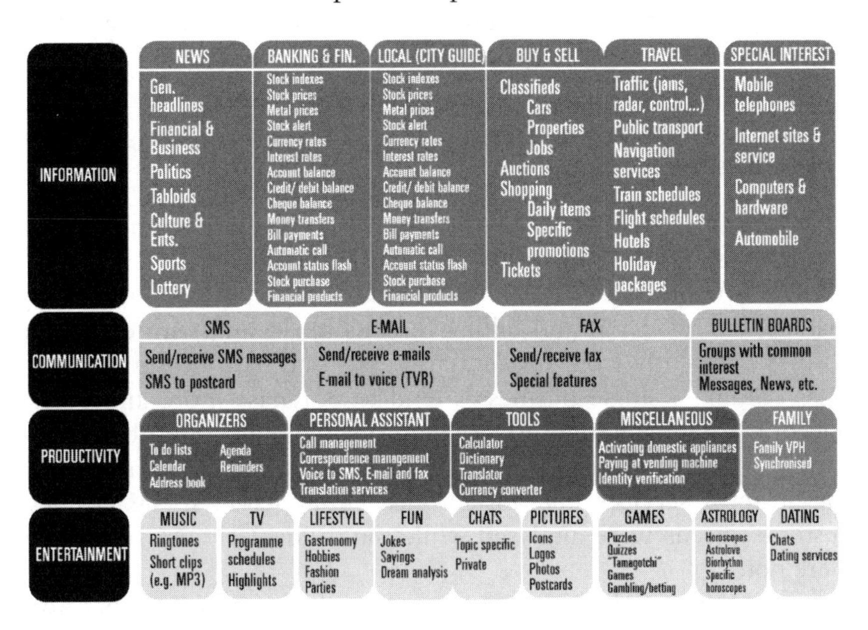

Figure 10.3 Hypermarket of services as defined by Nokia. Source: Nokia, Make Money with 3G Services White Paper.

As always, the customers will ultimately decide which outlets will flourish and which will die. The mega-market analogy provided here is to act as a mental stimulator, allowing for many lessons to be learned from the retail industry which has experienced similar issues in the past and found effective marketing methods to deal with them. It is not the only analogy and it will not provide the absolute truth. The analogy does, however, help illustrate to the other players in the UMTS environment, how great a change is taking place with the mobile operators, and perhaps help build some understanding of the challenges being faced by the organisations in this change.

10.3 Segmentation

Segmentation is the marketing process of dividing customers into groups called segments, which are homogeneous by need within any group and heterogeneous by need when compared against other groups. The ultimate purpose of segmentation is to find differentiating attributes for each of the segments which are relevant to that segment in the purchasing or consumption decisions, to help focus marketing efforts.

Telecoms segmentation today

Traditionally telecoms operators have divided their customers into a few groups, typically from two to five. A significant difference has been the split into business and residential users – partly mandated by history with taxation and billing reasons of dealing with corporate entities differently from residential users – and partly resulting from sales volume – single businesses tend to purchase more and bigger telecoms equipment and services than single families. Another significant reason in many cases was that since there was some competition for business services but not for residential services, the business customer needs were analysed while the residential customer needs were not.

Still in the mid 1990s in most countries the business 'competition' was the big dominant PTT (national Post, Telecoms and Telegraph organisation) offering all telecoms services and products to all customers, and small niche markets offered to business customers. These were typically PBXs (Private Branch eXchanges), cellular phones and services, datacom services and Internet access. Most early cellular offerings were not even attempting to offer full business solutions, only to cover the mobility part of a business customer's telecoms needs.

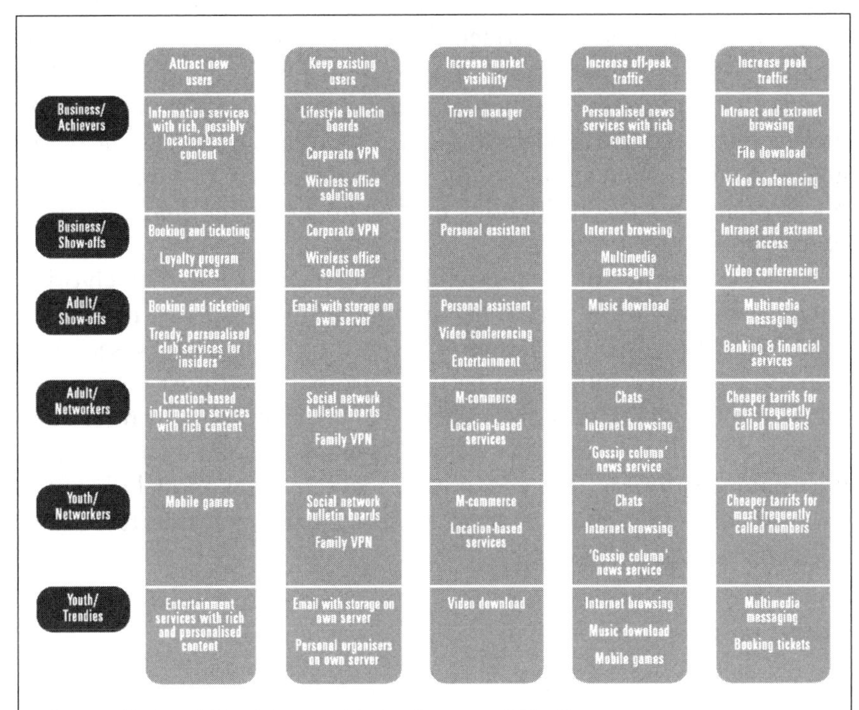

Source: Nokia, Make Money with 3G Services White Paper.
It is necessary to have a segmentation and services targeting plan to create the market for new services. This can help the marketing organisation and sales people focus on those segments that will exhibit the early desire for new services. Basic marketing skills and processes are just as relevant in UMTS as in other industries.

So from the operator's segmentation need, a single residential user bought little and had little or no competitive options, so the operator did not have to 'worry' about understanding too deeply the needs of their residential customer. But business customers spent more and were offered competitive services and products by private companies, and thus their needs had to be better understood. Partly for these reasons, most fixed operators even at the end of the 1990s had usually not more than four segments, where residential customers were one segment and business customers were divided for example into three as Corporate (biggest) customers, SME (Small and Medium Enterprise), and SOHO (Small Office Home Office).

As early mobile operators emulated and learned from the 'big' fixed operators, the mobile operators usually copied the fixed operators' simplistic segmentation model. Where the mobile operator was originally set up as a division of a fixed operator, this was also then the result of inheriting the parent's segmentation model. With many mobile operators the residential services focus has soon become stronger, so it is not atypical that the mobile operator arm of a fixed-mobile operator knows the residential user segment better than the fixed parent operator.

At the emergence of the pre-paid billing phenomena – which originated from a taxation related law in Italy and soon spread around the world due to customer desire – most mobile operators see their primary customer split being between pre-paid and post-paid customers, and usually have a preference to move customers from pre-pay to post-pay. In this case business customers are practically always post-paid customers. Note that this segmentation model is driven by needs of the mobile operator's *billing* department, not its marketing department.

Model based on customer need

A segment model's power depends on its ability to explain and predict behaviour of distinct groups of customers. If the model simply says that a large customer will spend more in absolute terms than a smaller customer, then we can question the benefit of the model. But when a model can take two customers of about the same size (or other such characteristics) and suggest why one customer will behave differently from the other, then the segmentation model will deliver value to marketing. And the more the model can isolate and identify unique motivators and causes for behaviour, the stronger the model will be as a tool to guide marketing efforts.

The size of the business is often proportional to the usage of communications services and on a simplistic level the total spend on a phone bill provides 'differences' for the operator, but size does not explain difference in behaviour and thus makes for very poor business segmentation models. For example a taxi service of 200 taxis and thus about 500 employees could be argued to be similar in size to a printing house of 500 employees. Yet a taxi service will place very many mobile phone calls with its drivers every day, while the print shop will have most of its 500 employees running the printing equipment and not needing to be contacted by mobile phone. Size is a poor predictor of mobile phone behaviour.

The type of business can be a much stronger predictor of the type of behaviour. The same taxi service of 200 taxis will be placing calls with its drivers continuously during the day. The calls tend to be local within the township or region where the taxis operate, and relatively short in duration to inform of the next job, but very many in a day. The dispersion of the calling traffic is relatively uniform amongst all drivers. The size of the company is not related to the proportional amount of mobile phone business traffic per mobile phone. A five-person taxi company behaves in a very similar way and has very similar mobile telecoms needs as a large SME taxi fleet company with 200 cabs and hundreds of drivers.

Both of these businesses would have dramatically different needs from the local printing shop which might very well make do with most of its telecoms traffic based on the fixed line and the internet, and the five person taxi service might place more mobile phone traffic than a 200 person print shop. And again, a small corner store printing shop with five employees will be more similar in usage to that printing chain with hundreds of employees. Size is not the biggest defining characteristic for mobile phone use in business segmentation.

Size does have some impact of course. For example small businesses tend to place most of their calls to numbers outside the company while very large corporations tend to have most of their telephone traffic remain inter-office, between employees of the company. But this behaviour pattern is more relevant to fixed telephony and services such as **VPN (Virtual Private Network)** which allows making calls between office location using short number 'internal dialing' etc.

Therefore any modern segmentation model needs to start from customer needs. With mobile telecommunications in the UMTS age, the customer need usually will involve two very distinct parts, the initial decision of *selecting* the UMTS *operator*, and the subsequent *usage* of UMTS *services*. The entities making these decisions can be very different at those two stages and will require separate customer analysis for either instance. To illustrate by example, a large corporate customer is likely to select its mobile service operator centrally, and make a bulk contract for thousands of users. Thus there is one central purchasing body which decides. But for actual services used from the service portfolio, a young travelling sales representative is likely to use a different set of mobile services than an older accounting department manager. We will provide a sample segmentation model focusing on the *usage* of services and explore that in more detail. Then we will briefly also discuss the segmentation of customers in the stage of *initial selection* of mobile operators.

The wide variety of possibilities enabled by UMTS mobile networks and the new business model for the whole industry make segmentation of customers even more challenging for the mobile operator. On what criteria does a customer select his or her mobile banking service? Is that the same criteria that the customer uses in selecting the news service? Traditionally there has been little scope for differentiation. The service offering for the whole customer base has been almost the same from one cellular operator to another. The main differentiation in many cases has been price with operators deploying various tariff plans for different customer types or by the hour of the day.

The dramatic increase in the overall number of services enabled by UMTS does not mean that the traditional split of customers to 'business' and 'consumer' would become obsolete but rather it becomes inadequate. Rather it can be seen that the basic split needs much further analysis into further segments, sub-segments and clusters. As was discussed earlier, the ultimate segment could even be a segment of one. This means that the ever increasing possibilities to personalise services makes it possible to handle all the subscribers individually as their own separate segments.

Being in a position to target individual users will require a degree of network sophistication, profiling and research. This will not happen at the start of UMTS service launch but will evolve as operators gain an increasing knowledge in their users behaviour and habits. So early on during the rush to introduce the first couple of hundred launch services each UMTS operator will segment the market at a general level. For these purposes a simple model is developed is discussed here to be used as an example.

10.4 Segmentation model on UMTS service usage

A two dimensional segmentation model has been used here to explain how users select and use individual UMTS services. Table 10.2. The first level consists of three groups: business, adult and youth. The second level segmentation divides each of the three segments into four further sub-segments, but for the purposes of example, we will examine only two of the four sub-segments for each of the three first level segments. It is important to recognise that in the UMTS environment, users will separately be selecting services to consume, which will be a different decision-making process from initially selecting the UMTS network operator (or MVNO (Mobile Virtual Network Operator) or service provider). This segmentation example looks at the segmentation related to consuming individual services. Later in this chapter we will also briefly look at the segmentation of selecting the network operator.

Highest level segments

If forced to divide the total mobile phone population into only three significantly large groups so that the initial division would have the greatest predictive impact, we divided the customers into Business, Adult and Youth segments. This division has meaning as these three groups behave very differently. As far as extremely simple models go, this model has considerable utility from its intuitive nature – we all know businesses, adults and young people, thus we all can immediately make generalisations about how any one of the three groups might behave differently from the other three. And this model does not need much study into social, psychological or marketing theories to be immediately usable.

Business users tend to have their employer pay for the cost of the call which alters their behaviour and greatly diminishes the threshold of placing a call. While many enterprises control telecoms costs of their employees, companies are willing to accept large telecoms

Table 10.2 Mapping opinion leader groups

1st level	Business		Adult		Youth	
2nd level	Performers	Show-offs	Show-offs	Networkers	Networkers	Trend-setters
Key characteristics	Business driven, heavy users, look for services that increase work efficiency	Significant business use, but the mobile terminal is important also for personal life. Being trendy is more important than improving business efficiency	Mainly personal use, like to impress others and to be trendy	Personal use, can stay ages on the phone, contact ability is very important	Under 25 years, personal use, can stay ages on the phone, being member of a contact network is valued	Under 25 years, not necessarily heavy users, but like to impress others by using fancy services
Adoption	In the 1st wave	In the 2nd wave	In the 2nd wave	In the 3rd wave	In the 3rd wave	In the 1st wave

Source: Nokia, Make Money with 3G Services White Paper.

expenses from at least selected groups of employees such as their travelling salesmen, etc. Business customers can quickly adopt a new technology if it is seen as producing a net (financial) benefit to the company.

Adult residential users pay for their own calls and are thus more price-sensitive per service use than business users. Adults have considerably more disposable income than the youth but adults tend to be less eager to experiment with the latest fads and fashions, so adults are usually later adopters of new technology than the Youth.

Youth users have usually some initial purchase and spending support from their parents, but soon will have to bear most of the cost of using services. Youth are eager to try new things and can be very 'irrational' in using a given service – meaning they can 'throw away their money' in what may seem like idiotic uses to their parents, but the Youth also has usually quite limited disposable income, so their total ability to consume is limited.

By population of customers, the Adult segment is biggest of the three, whereas by spending (and revenues to the operator) the Business customers spend the most. Of the three, Youth are the most eager to experiment and adopt new things.

Second level segments

The first level divided our subscriber base into three. If our UMTS operator had 12 million subscribers, each of the resulting segment would be typically millions in size. There is very little marketing utility in addressing a population which has millions. The differences are likely to be much more prevalent than the similarities.

To provide more utility from our model, we add a second dimension which has four groups. As with all multidimensional segmentation models, any dimension can be examined independently. This means that if used alone, this second level would also produce segments with populations in the millions. By adding dimensions we can subdivide segments into more precise but smaller segments, with ever more precisely defined needs. If the two segment definitions do not have a lot of overlap, then the resulting $3 \times 4 = 12$ segment matrix brings our imaginary 12 million subscriber population into 12 segments averaging a million subscribers each. A two-dimensional segmentation model for a million subscribers is still only the beginning, a meaningful model might have a dozen dimensions or more, producing 'micro-segments' often with populations only in the thousands or hundreds. Marketing theorists have suggested the ultimate segment is a segment of one.

Four subgroups

For end-users we have defined four groups of customers with distinct behaviour characteristics to isolate again a simplistic division of the population.

The types of individual end-user are: Performers, Show-Offs, Networkers and Trend-setters. Each Type of user in each segment has defined key characteristics that can be used for the targeting of initial services and to define if the user group is in the first wave or second wave in adoption.

Note that with a two-dimensional model, any given person will belong into one segment from the top level, and one segment from the second level. This means that someone could be both a youth person and a networker or another person could be both a business customer and performer.

Target users

The operator may consider targeting the users in different waves. To maximise the UMTS penetration the 'elite opinion leaders' within these segments need to be identified. 'Opinion leaders' are those who should first be impressed and attracted to use UMTS services. When they do this, they act as another marketing channel for the operator to promote UMTS to mass markets as they 'pull through' the followers. This is defined in Table 1. Three segments are identified first. Business, Adult and Youth. In each segment there are each of the four typical types of user, Performer, Show-Offs, Networkers and Trend-setters. In Table 2 we have shown only two of the four for the purpose of example and simplicity. Each Type of user in each segment has defined key characteristics that can be used for the targeting of initial services and to define it the user group is first wave of second wave in adoption.

At the beginning, the service offering can be targeted for *Business Performers* and *Youth Trend-setters* in order to raise interest in UMTS mobile services. Business Performers and Youth Trend-setters are willing to be at the forefront, want to be seen using new services and are willing to adopt new technologies.

For the *Business Performers*, PIM (Personal Information Management) such as *e-mail, messaging, address book, calendar* will be important to manage their profile as well as *personalised news services* with rich content. *Video conferencing* through their mobile terminal will be well suited for these business-driven, heavy users who are looking for services that increase work efficiency.

The *Youth Trend-setters*, whose aim is to impress their peers with fancy new services, can be induced with mobile entertainment services such as *mobile games, multimedia messaging (including e-mail) and music downloading*.

After the initial phase of the UMTS life cycle, it is time to start progressively paying more attention to other opinion leader segments and at the end of the day to the general public as well. At the same time it is good to notice that from the operator's viewpoint, the multitude of UMTS services will serve different purposes. At least the following business drivers for services can be identified – at the end of the day they all target to increase operator's market capitalisation:

Stimulate growth by capturing new users

When considering the business case for UMTS this means offering services that encourage customers to start subscribing to UMTS services. In practice this means delivering services – or service bundles with services – that cannot be provided without capabilities of a UMTS networks (i.e. cannot be provided on 2G networks in the same way and in similar format).

Increase market share by reducing churn

Some services should be targeted to decrease churn and 'lock-in' the UMTS subscriber base. An e-mail service with the inbox on the operator's server is an example of such a 'sticky' service that makes end-users reluctant to switch service providers. Note that churn is managed primarily by having the user customise their messaging service (compared to e-mail address switching). Churn management applications are typically user content (person-to-person) applications (and segment generic) rather than commercial content oriented.

The operator must also know which customers are the most likely to leave the operator. Customer intelligence and segmentation have an important role in this too. Micro-segments which have high churn rates require new solutions to stop the leaks and/or solutions to win back customers. Customers whose characteristics are similar to those identified as probable churners require special care from the operator. Churn is discussed more in the Competitiveness chapter.

Create interest by enhancing market visibility

Creating interest – even hype – is needed to market UMTS services. Some services should be such that they raise the interest levels of the general public. Video-telephony, Music Downloads and Games could be examples of these types of high interest services. The degree of interest is also segment specific – for example: video-telephony for Adults, Java based-games for the Trend-setters, calling-line pictures and music download for the Show-offs.

Increase profit by encouraging off-peak traffic

Intensifying off-peak traffic will have a direct impact on the operator's profitability as it increases utilisation of the UMTS investment. Delivering digital 'newspapers' to subscribers' terminals during the night when there excess capacity, is an example of the services falling into this category. Newspapers are one emerging application but some other night time activities (e.g. promo music videos, upload/download of device/car information, video surveillance, etc.) are also envisaged.

Generate cash flow by intensifying peak traffic

Peak traffic, which is typically priced at a higher tariff, is important to generate cash flow. UMTS offers the opportunity for operators to tariff services based on QoS (Quality of Service) – where best effort transmission could be priced at a fixed tariff to win new users and sell higher QoS to those users with higher requirements. The Personal Assistant type of work related productivity applications are likely to increase peak traffic.

By achieving all of these objectives the operator will be able to increase its market share providing it can create services that also satisfy user needs. Despite this common goal, sub-objectives should be prioritised for each phase of the UMTS life cycle since all services can not be created and introduced at the same time. This will enable the operator's service creation managers to prioritise and select the applications and services to be offered for customers. Definition of concrete service offering needs to be done by mapping together the operator's business objectives and the target segments. Table 10.3 illustrates some suggested main applications to achieve these targets.

It should be noted that in Table 10.3, the six segments described represent only the key opinion leaders. This by no means represent the entire potential UMTS customer base. The underlying assumption is that the rest of the mass market will follow the lead established by these key groups.

Another assumption has been that the same application can have different impacts, depending on the segment in which it used. On the other hand, if the service has an impact on more that one of the operator's business drivers, then the service is only mentioned in one column of the table, (i.e. that column where it is estimated to have the biggest impact).

10.5 New segmentation methods

Note that the above model is still an oversimplified model which can serve in areas of limited competition and emerging markets, but the above model will be of almost no practical use in real UMTS operator environments due to its oversimplified nature. The same type of thinking needs to be developed several iterations further, identifying attributes that can be identified to divide the subscriber population further. A multidimensional model gets ever more cumbersome to construct and to use the more dimensions are added to it. Many other techniques exist and are being developed for more powerful segmentation.

Table 10.3 Key priority applications during the early phases of the UMTS launch

	Attract new users	Keep existing users	Increase market visibility	Increase off-peak traffic	Increase peak traffic
Business/ Performers	Information services with rich, possibly location-based content	Lifestyle bulletin boards Corporate VPN Wireless office solutions	Travel manager	Personalised news services with rich content	Intranet and extranet browsing File download Video conferencing
Business/ Show-offs	Booking and ticketing Loyalty program services	Corporate VPN Wireless office solutions	Personal assistant	Internet browsing Multimedia messaging	Intranet and extranet access Video conferencing
Adult/ Show-offs	Booking and ticketing Trendy, personalised club services for 'insiders'	Email with storage on own server	Personal assistant Video conferencing Entertainment	Music download	Multimedia messaging Banking & financial services
Adult/ Networkers	Location-based information services with rich content	Social network bulletin boards Family VPN	M-commerce Location-based services	Chats Internet browsing 'Gossip column' news service	Cheaper tariffs for most frequently called numbers
Youth/ Networkers	Mobile games	Social network bulletin boards Family VPN	M-commerce Location-based services	Chats Internet browsing 'Gossip column' news service	Cheaper tariffs for most frequently called numbers
Youth/ Trend-setters	Entertainment services with rich and personalised content	Email with storage on own server Personal organisers on own server	Video download	Internet browsing Music download Mobile games	Multimedia messaging Booking and ticketing

The UMTS operator needs segmentation models that are deeper, more accurate and more appropriate for that particular operator's strategy and customers. In order to have a successful and profitable customer relationship the operator must split customers to many segments and micro-segments and be able to personalise the offerings and services. It is also important that the customer split into segments is done based on real data, not only traditional models and their assumptions.

The operators are able to collect and generate data, for example, from the billing system, CRM (Customer Relationship Management) and mobile portal. New data analysis technologies, such as neural network models, open new opportunities to real customer intelligence and effective segmentation. The ultimate target must be to utilise this information to develop services more attractive and valuable for users and not only for advertising, which can spoil the customer relationship.

For existing operators the transfer from a network centric model to a customer centric model is often one of the main challenges that arise from UMTS. The customer centric model often requires dramatic overhauls of the operator organisations, processes and IT (Information Technology) systems. New tools, such as data analysis systems based on artificial intelligence, require processes which can utilise information and micro-segments. This transfer is not easy for the operators, but it is necessary in order to survive in the fierce competition in the future.

10.6 Segmentation of initial network operator selection

In addition to the daily choices we face in selecting individual services on our UMTS phone, we will also be occasionally in the position to select or change UMTS network operators. On what criteria does a customer select his or her mobile operator? In second generation services the service offering has been almost the same from one cellular operator to another. The main differentiation in many cases has been price with operators deploying various tariff plans for different customer types or by the hour of the day.

The customers select their service provider when they sign up for a service, and may change (churn) to another operator from time to time. Operator selection is unlikely to happen more than a few times per year, and many customers will select one operator and stay with that service for years. Increasingly customers will also have accounts with two or even three operators and use them according to price or service preferences.

With business customers the initial provider decision is usually made centrally by the company, and individual employees do not select operators. For the residential users the decision to use a service provider is often made by the head of the household, at least for the first mobile phone, or by the user who is most familiar with mobile phones for example from using one at work. The youth may have their first mobile phone and operator selected by the parents for family package benefits and discounted costs of calls to the parent's phone. The Youth is often extremely influenced by peer pressure and which type of phone or service is the 'in' thing to have or what is the 'coolest'.

This book is not about operator strategies on building customer bases but rather on Services for UMTS and how to market those. The above segmentation has focused on how to use segmentation to increase usage in a focused way. But before people can increase usage of UMTS services, they have to sign up to them. That is why a brief discussion was included to cover the aspects of signing on new users. A thorough study and understanding of why customers select operators is of course needed to build a solid segmentation model for this purpose as well.

10.7 The UMTS operator brand

All brands make promises and offer a guarantee of satisfaction. A brand needs to make users feel an emotional attachment that comes about because they feel the service is more suitable for them than anything offered by any competing service. Brands must engender trust and loyalty if the products or services they represent are ultimately to be purchased.

15 Vignettes from a 3G Future

Instant video conference

I remember how difficult setting up video conferencing used to be. There were special long phone numbers to dial and we used custom equipment. Now it has become very common that if someone is suddenly unable to make a meeting, we have them join us via video conferencing. Most phones have the cameras and now most of our conference rooms also have large screen monitors to display a larger image of the small-screen picture of the people from the other end. In an emergency we connect the UMTS phone to a laptop computer to get a bigger view of our meeting counterparts. We find that some of the perennially late people have taken to the habit of joining the beginning of the meeting via video conference while still in the taxi.

Video conferencing will be a very commonplace ability of UMTS not unlike SMS is standard in GSM, or attachments commonplace in e-mail. The costs of real time live video will of course be considerably more than the costs of voice calls, but again the costs and convenience will be much better than current custom solutions using ISDN etc. Video conferencing will become a common means to participate in meetings where travel costs or schedules would make it otherwise impossible to meet. Operators will be pushing video conferencing very strongly to their business customers and probably offer various enhancements to business video conferencing, such as Bluetooth or W-LAN based connections to large screen displays and video projectors, so that the video conference participant can be seen clearly by the participants of the meeting.

Initially UMTS operators will be launching some hundreds of services out of a potential service universe in the several thousands of possible services. In any given market during the first few years the UMTS operators will have distinct services which differ from the competition simply because every player has not had time to replicate every service offered by the competition. As time goes by, more and more of the popular and profitable services will be copied by the competition, and thus the UMTS service offering by all players start to become more and more similar. At that stage branding will become one of the keys, if not the key, to market leadership – where customers will prefer one service over the other, partly or mostly, because of its brand-image.

That brand power cannot be created at that time when it is really needed. It has to be created now, well before it becomes critical for success in the marketplace. That is why UMTS operators have to build a strong brand strategy from the start, and ensure that all marketing activities – including positioning in the market, handling concerns and problems, and messages delivered to the preferred target segments – are in full support of the branding strategy. Brands are built over years and decades, not over months. Brands transcend the mere services side of UMTS, and we recommend that operators and content providers get familiar with current thinking on branding strategies to ensure the value benefits of branding.

Multiple brands

The brand strategy does not mean that the UMTS operator's brand will be the only brand in the portal or channel. Just as there are many product brands on the shelves of the mega-markets, there will be international brands and regional and local brands visible in the operator's portal. There will also be a variety of brands in the distribution channel. Mobile operators today rely both on their own, wholly owned distribution chains as well as independent distribution chains. Both will also exist in the UMTS world.

Again we can use the retailer model as an example of how the UMTS services offering will work. The fruit and vegetables product offering is normally branded by the retailer since there is no one brand in this area. There is segmentation by quality for instance with organically grown produce and this can become a differentiating offer from the retailer. It is a similar case with meat and fish, bread and other basic food stuffs. Again there can be differentiation in these product offers between retailers if one is focusing on price and the other on higher quality produce. Regional brands can be apparent in this area, Argentinean or Scottish beef for example that is marketed as higher quality. These areas are where the retailer can build their own brand image around the produce offer and for the UMTS operator this equates to services that they can develop in house or with content providers who do not have a brand of their own.

In other areas of the store international brands are also visible. Coca-Cola, Pepsi in the soft drinks section, Coors, Budweiser, Heineken and others in the alcohol section, Campbell's and Heinz in the soups section. All these brands have their own position in the market and the food retailer does not try to replace them, only to enhance them as part of its total customer offer.

A clear example of this is financial services. If the UMTS operator decides to become a bank it will be competing against existing banks who often have a strong brand and market position. Most customers will already have their own bank and in the main will be happy with the service. To displace this loyalty will be difficult especially in the short term so it is necessary to have service agreements with most if not all of the banks and it will be the bank's brand that will be most prominent in this service. So the UMTS operator also has to become a content aggregator and use a variety of brands as part of their total service offer.

10.8 Loyalty schemes

UMTS technology will give operators the ability to offer great services via exciting new mobile devices, which have the potential to do just that. An example of these loyalty generating services can for example be in situations where instead of carrying mega-market and airline loyalty cards a mobile operator could team up with a mega-market and an airline to offer rewards to end users. The operator could store all the details in the users profile for continual use.

With differentiation possibilities about to explode the brand strategy will become ever more important – it will become possible to enhance the reputation that operators enjoy in the minds of end users thereby encouraging greater customer loyalty.

Customer retention strategies

Experience has shown that those customers who are successfully targeted with 'sticky' services remain customers for longer. The introduction of UMTS services should contribute substantially to churn reduction if operators 'get it right'. The experience in Japan where i-mode services have taken the market by storm is that churn rates are running at around 1%. There are many initiatives undertaken by operators to reduce the impact of churn on their business.

Customer reward programmes

There are many customer loyalty programmes already in existence in many industries, and are used with varying degrees of success. In all cases they need to offer a very strong incentive for the end user to stay with their current supplier. There are few telecommunications operators who are currently using loyalty programs effectively. A good UMTS loyalty program will offer substantial rewards – if possible linked to handset upgrade which is often a reason for churn. With the launch of UMTS terminals the operators who are late to the market with services may see increased churn by users who want exciting new terminals and services now and are not prepared to wait.

UMTS operators need to consider the whole spectrum of customer loyalty programs that currently exist in the market and should take the best components from those programs that fit to the UMTS business model and their own customer loyalty strategy. Some ideas could include:

- UMTS loyalty points for usage similar to airline loyalty schemes.
- Different membership levels with the Gold, Silver or Bronze loyalty cards stored in the UMTS terminal entitling users to special benefits like priority notice of special events organised by the Operator or major sponsors.

- Extra points for every year that users are registered to reduce churn; users could earn double points after 2 years loyalty or get one-off bonuses each calendar year.
- Points earned for playing on-line games with the winners getting extra bonus points.
- Special promotions targeted to the higher usage Gold members like additional discounts from certain retail outlets.
- Joint promotions with major sponsors towards the members that are designed to build the brands of both UMTS operator and the sponsors.
- Priority customer care access so that Gold members wait for less time when calling the care line.
- 'Gold Cross' days when Gold members get a percentage discount on specified UMTS services, similar to 'Discount Sales' in the retail world.
- Gold members receive prior notice and usage of new UMTS services – first to try options again with pre-launch discounts.

These are just a few ideas but there are many that can and will be found to encourage loyalty and usage in the UMTS business.

CRM – Customer Relationship Management

CRM has been discussed elsewhere as a service which the operator can provide for its corporate customers to use with mobile services. CRM also works of course for the UMTS operator itself where an operator can really make the customer feel like they are 'wanted'. By making good use of the large amount of customer data the operator can begin to create an understanding of its customers and start to build their emotional attachments to the operator. The smart operator can tailor marketing programmes for the individual – or at least for specific segments. Customer care can be prioritised and linked to the Customer Loyalty program. CRM was discussed in more detail in the Movement Chapter.

10.9 Don't forget the UMTS distribution channel

Probably the biggest challenge for the UMTS operator will be the limitations of the distribution channel. Most mobile service purchases today revolve around the terminal and that is unlikely to change with UMTS. What is different is that the mobile phone retailers will need to now sell services. This means moving from selling boxes to selling access to content that has value. Unfortunately many of the current retailers are not geared up for this. There is a variety of outlets from specialised mobile phone centres to large electrical retailers who sell mobile phones alongside DVDs (Digital Video Discs), cameras, computers, etc. and some will find it harder than others to handle the change to selling UMTS services as part of the phone package.

The UMTS operator needs to do two things. Provide extensive training into the channel and make is simple to sell new services. The benefits from training are medium to long term while the benefits from simple service packages are more immediate. In fact we would say that without the short term benefits of simple service offerings the long term growth will be in jeopardy.

Keep it simple stupid!

In the short term we believe that operators have to live with the service selling limitations in the channel. So the UMTS offer has to be packaged in a way that makes it simple for the mobile phone retailer to sell. If the predominant capability is box selling then this is the initial marketing strategy that needs to be adopted. There needs to be a service box that the retailer can display and sell. There is also a need to appreciate that although customers are becoming more mobile aware and increasingly more sophisticated in their demands and tastes, there is still a limit to their capability to understand what can be a complicated sales pitch. As we will discuss in the next chapter on Competitiveness it is difficult for most customers to appreciate what 1 MB of data really means in any quantifiable way. A deli offering an 'all you can eat service' for $10 is similar to a mobile operator offering 400 min of voice calls for $40 per month since both can be visualised. Not so with bits of data. The

UMTS services have to be packaged and marketed and sold in terms that the consumer fully understands. In other words the offer has to be quantifiable.

Packaging (bundling) UMTS services

If the retailers are good at selling boxes then one option is to give them a box to sell. As it turns out this can also be one way to initially market to defined segments of the customer base.

Sports fans could be one identified segment and these customers are generally interested in sport news and content. This opens up the possibility of a '**sports package**' service that could be along the lines of 20 sports news stories per month for $5. Or a specific service based say around the Olympic games, football world cup or Wimbledon tennis competition. The service could be 'packaged' in a box along with the UMTS terminal, branded with the operators logo with images of various sports on the box. This also opens up the possibilities of sports personality or team endorsements to further enhance the 'brand'. This kind of approach can be extended to Music and Games for instance with the Music Service package boxed with a UMTS terminal that has MP3 and FM radio capabilities and the Games package boxed with a new UMTS gaming terminal that has 3D interactive gaming features.

To make the service simple to customise, the UMTS terminal when activated can point the customer towards the Sports content page in the case of the Sports service for example where the types of sports news; soccer, ice hockey, basketball, etc. can be selected as areas of interest. Again this is removing the need for the retail outlet to know how to offer and sell personalised content. The incentive for the retailer can be a simple extra commission for selling a UMTS package instead of just a terminal.

Since most sports, music or games enthusiasts will not just want this one service the boxed service can include promotions of other services like general news content, what's on? and in fact any of the services outlined in this book and more. There only needs to be the marketing literature provided to create the awareness of what is on offer and the bookmarks in the UMTS terminal to make access, subscribing, and personalisation simple.

16 Vignettes from a 3G Future:

Sports Events Anywhere

I love my Formula One racing. But sometimes I am travelling on a Sunday and miss the race. It used to be that if the local TV did not carry the race, I would be desperate to find some coverage of the results. Now I have the ability to use the small screen and UMTS access to get to view the start of the race live, no matter where I am. I could watch the whole race live on the phone, of course, but watching TV is still quite expensive on UMTS. Still I love it that I can watch the live race for a while and then switch to the text-based real time race updates. It is not quite the same as watching the race on real TV, but where that is not possible, this is by far the next best thing. And the cost of a couple of minutes of the race is not really that expensive.

UMTS Operators will probably partner with major sports events owners to allow UMTS terminal viewers premium cost viewing of the various sports events. Of course the small screen and extra cost will mean that if regular TV is available, that will be the preferred means of viewing. But there will be huge audiences of interested fans who happen to be beyond the TV coverage of any given sporting event. Many of these will be willing to pay something for seeing a part or even the whole event. Various update services are likely to emerge giving real-time coverage of still images and event updates as appropriate.

Service packaging in a 'boxed' solution can be the first step to creating the UMTS services market and growing usage. As the knowledge and experience in the channel grows it becomes easier to offer more complex services through owned and independent distributors which is where the channel training then start to have a positive impact on an operators bottom line. After all that is what we are trying to achieve is it not?

10.10 Preparing for launch

When preparing for launch a structured and well-planned approach is vital. Apart from well-prepared segmentation, operators need to cover several other launch issues. Basic elements of the launch plan are:

Segmentation plan

A segmentation plan will define the early market adopters and split them into a number of categories where specific services can be targeted to maximise the take up and usage. Much of this has been discussed already in this chapter.

Marketing plan

A standard marketing plan covering at least the product/service offering, bundling, pricing, distribution chain, market and competitor intelligence, promotion, advertising, sales support, sales, and delivery. The plan includes of course budgets, resources and schedules.

Organisation transformation

The UMTS operator needs to transfer its organisation to become more customer centric and to be able to manage short cycles in developing new business models and services. Beyond succeeding with the more challenging customer relationship the operator must also be able to manage thousands of partners and application developers. The operator must launch new services all the time, be able to react to market needs and find the available applications. For the organisation this means flexible requirements, to be open and able to adapt. However, this also requires effective management to control profitability. The operator does not only launch new services all the time, it must also be able to discontinue products if they are not profitable or have failed in the marketplace. The partnership aspects of UMTS are covered later in this book in the Partnerships chapter.

Service definition

The operator needs to determine a service creation strategy and produce a definition for selected services – or service bundles – together with a proposal for a full marketing study. The service creation should be made to be very easy to develop, deploy, manage, maintain and extend services. Very important to keep in mind in UMTS, is that while speed in initial roll-out of a new service is important, *more important* is the speed of adapting to new market conditions. So the services created, and the platforms which are used to manage services, need to be very flexible and allow for rapid changes.

OSS analysis

The operator needs to analyse the OSS (Operational Support System) environment for the implementation of the operator's chosen services. Some of the systems which need to be analysed include Customer Care and Billing, Mediation and Rating, General Ledger, Interconnect Billing, Credit Card/Checking and Banking, Fraud Management and Network Management Systems. While most existing mobile operators have these, they are usually not prepared to handle money flows in multiple directions – to include revenue sharing with content partners, etc., and are not prepared to handle outside partners and their IT systems for example.

Application support

This phase involves the implementation of a selected number of applications and the introduction of specific IP (Internet Protocol)-based services. Most modern services would include various Client/Server systems. The objective of this process is to give the operator an understanding of what is required to deploy applications within the network, the flow of information from these applications and their billing implications. The system needs to be designed to handle the latency and various delays when individual servers query each other on solutions which are a combination of several independent services – such as a sponsored gaming site for multi-user games.

IP planning and deployment

This phase aims to provide the operator with guidance and hands on experience of deploying Internet and Intranet interconnections with the UMTS core network. The operator will need to consider mirroring selected content and applications to the edges of the network, closer to the users, where the delays would make the UMTS service otherwise impractical or annoyingly slow for the end-user.

Handling churn

Churn means customers leaving the company to become customers of a competitor. Understanding churn is a vital part of success in marketing. When mobile subscriber rates have grown at high speeds, it is perhaps understandable that operators have focused on customer acquisition. One should remember, however, that the cost to keep a current customer is a fraction of that to acquire a new one. Now the operator has to develop its understanding of its customers and adjust its processes and systems to respond to customer needs. We will examine Churn in more detail in the next chapter on Competitiveness.

10.11 Marking off marketing

This chapter has set the stage by examining the retail services model and applied lessons to UMTS. The chapter looked at how UMTS services should be marketed using standard theories of segmentation, deploying launch services with targeted and prioritised methods while taking into account existing channel capabilities so that they can create customer value and good returns on their investments.

Operators must remember to cover all of the basics of their repertoire and increasingly churn management and loyalty schemes will become a factor in the competition in the UMTS environment.

In deciding which services to create, the UMTS operator should not try to create the perfect single service, but rather keep experimenting with lots of ideas that seem possible, to find the most successful services. As the hockey goal-scoring legend Wayne Gretzky said: "You miss 100% of the shots you never take."

11

'When written in Chinese the word crisis is composed of two characters. One represents danger and the other represents opportunity.'
John F Kennedy

Competitiveness in UMTS:
The Winner Takes it All

Joe Barrett, Ari Lehtoranta, Canice McKee, Jouko Ahvenainen and *Tomi T Ahonen*

The competitiveness of any market is a factor of numerous strengths, weakness, threats and opportunities that exist within the market place, and intrinsically within the operator's organisation itself. The competitive environment changes over time. Entry barriers come and go, expansion and consolidation create new market situations, economic conditions change the foundations that yesterday looked solid and firm. Successful companies are those that can continually mould their workforce into the best fit for the prevailing market conditions and more importantly can predict what those market conditions could become.

Competitiveness is not always about new products and being first to market. History is littered with companies that created a revolutionary new product but failed to capitalise on that advantage. In UMTS (Universal Mobile Telecommunications System) it will be no different. The key is to know your market, know your competitors and most of all, know your customers. In this chapter we will focus mostly on the importance of customer knowledge and how that can be one key differentiator in gaining a competitive advantage.

11.1 Operator vs non-operator

In UMTS the operators are not alone in the business anymore. The operator's environment has changed, new players have emerged, and the operator itself is exposed to many new opportunities and threats. The value chain is becoming more complex. As we can see in Figure 11.1 the value chain offers many different roles for the operator. The operator needs new competence areas, but it cannot handle the whole new value chain alone. The UMTS operators must choose its own role, focus on it and find top level partners to fulfil missing pieces of the value chain.

As operators evolve their business's from the delivery of voice with limited mobile data services to providing a complex mobile services offering, they will be facing numerous challenges. This will come from new types of competitors who in some cases may also be partners, i.e. numerous application developers, content providers,

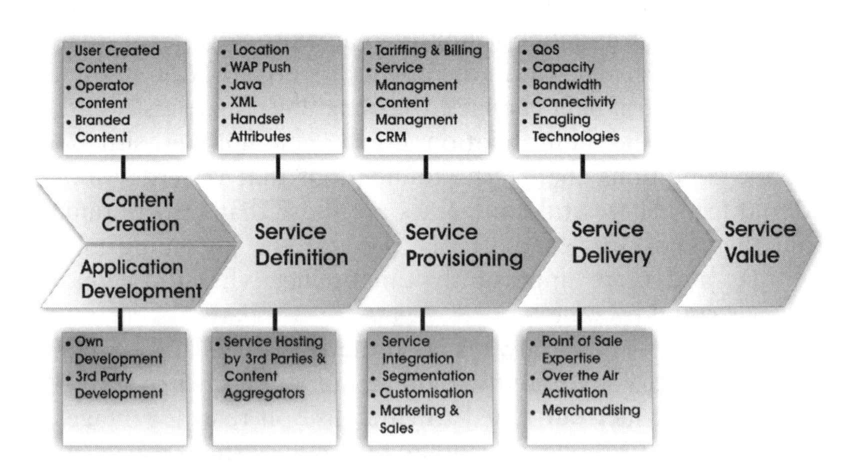

Figure 11.1 The UMTS business value chain.
Source: Nokia.

MVNOs (Mobile Virtual Network Operators), IT (Information Technology) integrators, etc. The UMTS network operator has several key competences and competitive advantages that place it in a unique position to exploit new revenue opportunities.

Operator competitive advantages

It should be mentioned that operators have several key competences and unique competitive advantages. Both the operators themselves, and their prospective partners need to understand what these are and how exclusive they are.

Control of network

The first, and most obvious competitive advantage is control of the UMTS network. The network and its various resources, including exclusive rights to parts of the radio bandwidth, are a resource which nobody other than other UMTS operators can match, and nobody currently can exceed. The UMTS network is the most cost-effective wireless, wide area telecoms and data distribution system and its control yields considerable competitive advantages over any other medium and resource. While other technologies may come close to or even exceed the UMTS network in some technical areas like capacity or data rates, none can meet the cost efficiencies of providing continuous voice and data services to the mass market.

Billing

The second significant competitive advantage for the UMTS operator is the billing system and billing interface. No other industry tracks and bills for items of such small value, at precise intervals and quantities as seen in the telecoms industry. The billing systems deployed by Internet operators and MVNO's are almost definitely going to be less sophisticated and provide less data than the systems deployed by UMTS operators.

User information

A third obvious competitive advantage is the control of the information on the caller especially as more and more services are delivered to individual customers. Even with pre-paid callers, the operator has a great deal of control from knowing the card number and card behaviour, even if the exact customer details are less than with post-paid billing situations. The larger ISPs (Internet Service Providers) may claim that they also have detailed information on their customers and have been tracking usage and behaviour patterns for some time. Yet we would argue that what they are tracking for most of the time is usage of a stationary PC (Personal Computer) which could have a number of users in the household.

Voice included

UMTS operators have an exceptional ability to deliver voice as in telecoms calls, and an almost exclusive ability to deliver mobile voice calls. This ability is an asset that can be leveraged. An ISP may attempt to bundle services around its ISP portal, but only mobile operators can bundle services around the mobile phone voice call service. Most users do not carry their Internet portal with them on a daily basis, even if they have a laptop computer. But after acquiring a mobile phone most users insist on carrying it with them everywhere. This is an incredibly powerful aspect of the mobile phone customer that no other media can deliver. People do not insist on carrying their colour TV with them everywhere, nor carry their used newspaper all day with them, nor haul their laptop computers everywhere. Yet mobile phones are carried almost everywhere. The foremost reason why that is done is the ability to make and receive mobile voice calls.

Location information

Another significant benefit is the location information of the caller/user. This advantage has been discussed in considerable detail in Attributes, the 5 M's (Movement Moment Me Money Machines) and the Movement chapter.

Brand and footprint

UMTS operators have national recognition and a national footprint. In terms of comparing available advertising and content delivery media, few are as solidly national in footprint as the UMTS network operator. Most mobile operators build network footprints to cover 80–90% and beyond within a few years of service launch. No city-based newspapers are able to deliver this type of national footprint. Many TV stations have regional coverage, and many radio stations do not provide national coverage. Magazines may provide national coverage but with a small and fragmented readerships. Now as many developed countries are reaching penetration levels in the 60, 70 and 80% points these operators have a national audience, a unique delivery method, and the back office systems to offer an abundance of new customer experiences. The only remaining question is, 'How is it going to happen?'

11.2 Operator vs MVNO

The MVNO will be a new phenomenon in most countries at the advent of UMTS. MVNO's will be marketing mobile services but will not own radio network assets nor licenses to use the radio spectrum. So MVNO's will need to lease capacity from one of the real network operators. The MVNO's ability to compete in the UMTS marketplace will be somewhat hindered by the fact that it does not control the infrastructure and will need to specify what it wants from its network operator partner through the partnership agreement(s). We will discuss MVNO's in more detail in the next chapter on Partnerships. In most cases to the end-user, an MVNO will seem just like any of the network operators and the rules of the marketplace will apply to competing with an MVNO very much like competing with another network operator.

17 Vignettes from a 3G Future

Pay my Bills When I Want To

One of the best things I love about my UMTS phone is the mobile banking. I still remember the time when I had to go to the bank teller. Then came the cash machines and telephone banking. Eventually it became possible for me to pay my bills via the ATM machines, and Internet banking emerged. But now I can do all of my bills and money transfers with the press of a few buttons on my UMTS phone. What is best, I can even do it while I am travelling in another country. I can even "wire" money to a nearby cash machine and take out cash in the local currency.

Banks and UMTS banking services can be built to take advantage of the personal nature of the UMTS terminal. Still most banking firms would probably like some kind of personal security in the forms of an PIN code or something else to guarantee that a borrowed phone or stolen phone is not used to empty someone's bank account. But the convenience of mobile Internet banking will enable a lot of digital electronic personal transactions by banking consumers. Operators will be in a perfect position to take advantage of these service opportunities, and traveller-oriented banking services will become early areas for service differentiation.

11.3 Operator vs operator

Most of the early interest in the telecoms competition will be in the market share battles between network operators. There are several likely scenarios that will be played out in most markets. There may be one or two established operators that do not own a UMTS license and will fight for market share on a 2.5G network. These will try to convince their customers and the marketplace that their technology is as good if not better than UMTS.

In each market typically there are one or two new entrant UMTS operators called 'Greenfield' operators which have not been operating a network previously in that country. They will be eager to show the best possible benefits from the new technology and attempt to profile themselves as the clear leaders in this area as they have no legacy networks to worry about. There will, of course, be a few established network operators who will have customers on 2G and 2.5G networks and will need to consider how to migrate these customers while fighting the challenges from the other players and keeping a credible story about their role in the transition of technology. Any MVNO in the marketplace might take on the role of any of the three competitors mentioned above.

11.4 UMTS operator vs 2.5G operator

When we look at a typical new service idea as expressed almost anywhere in this book, any telecoms engineer involved in GPRS (General Packet Radio System) or other 2.5G technologies will probably say: "I can do that with my technology today". Except for a few very high bandwidth services it is mostly true that almost all services in this book can be built using GPRS and a service could easily seem the same to the end-user. With UMTS the competitive advantage when compared with 2G and 2.5G networks does not come from a vast array of new service options, but from capacity and from the cost of delivering that capacity.

Capacity analogy: airline Industry

To draw a simple analogy from the airline industry – most jet airliners today travel at about the same speed and can offer a single passenger the options of a similar service. An airline can outfit a Boeing 707 or McDonnell Douglas DC 8 to have modern first class seats which recline to full beds, and Business Class seats with fancy video screen entertainment, and typical Tourist Class seats in somewhat cramped seating arrangements, etc.

For a single passenger the travel experience could be made to be quite similar to that on the latest wide-body jet by Boeing or Airbus, in fact if the airline wanted to, they could mostly install precisely the identical seats and entertainment, etc., although there might be severe indoor layout penalties trying to fit the most modern seats into narrow body airplane frames which are 40 years old. The flying speed is very similar, and in most cases the flying range is also similar, meaning that for a passenger experience, the first generation jet liners could theoretically provide an almost identical experience as the very latest ones.

The issue is not airliner speed, nor is it innovation in seating design. The issue is capacity. The current generation wide-body jets can carry many more passengers at much less cost than the first generation passenger jets. Many technological innovations have helped in creating the better cost-efficiency, certainly two of the most important aspects are the interior design of 'wide body' where 50 or 100% more passengers can be carried on a plane of roughly the same length and wingspan; and more efficient engines which can move a much bigger payload than the early jet engines on the first generation jet airliners. If we have on average 1000 paying passengers to fly on our airline from London to New York, it becomes an overwhelmingly superior business case to fly them in three modern wide body jet flights rather than in six of the jets of the first generation. It is a question of cost efficiency in handling high capacity.

When we compare 2.5G and 3G (UMTS), we see a similar efficiency issue. If you want to provide a data connection for a very small customer base and they do not transmit too much data on a daily basis, then current network technologies can handle the traffic load on a somewhat competitive cost. But when the mass market happens and we migrate more and more of our various information, entertainment, business and communication activity to the mobile terminal, the traffic load will grow exponentially and very soon capacity is the main competitive issue. At that point any operator who is trying to compete with an offer based solely on a 2.5G network will find itself in an ever-worsening cost-efficiency deficit. The cost efficiencies that come with UMTS mean these operators can gain market dominance by passing their lower costs onto the end-users.

Capacity is starting to be an issue today

Capacity in second generation networks is already approaching its limits in some markets and conditions. We see several examples of capacity constraints where the current 2G network is operating at much beyond its originally designed operational efficiency levels. In many mobile developed countries like the UK, Germany, Sweden, Finland and Italy where penetration rates are over 60% and in some cases over 80%, capacity is becoming an issue. In some major urban areas like London or Paris the likelihood of a dropped call while moving is fairly high. In other cities the indoor coverage is very weak reducing the ability to make a call. Network quality is stretched to the limit in many cell sites with no ability to increase the capacity due to the fact that operators have used up all their spectrum. One option is to cell split and install more micro-cells to increase the capacity density. However in dense urban areas finding suitable sites is a problem and can be very expensive. UMTS is needed in these cases for capacity relief.

Furthermore, as penetration levels reach saturation and the rate of new subscriber additions slows down, the focus for mobile operators becomes customer retention not customer acquisition. By default this means more focus towards the quality of the network and providing differentiating services that make it less likely for customers to churn to other networks. If we also assume that GPRS does take off and creates increased data volume there will be additional pressure on capacity that will impact network quality and the user experience of voice and data services. UMTS will solve both these problems with the capacity from new frequencies and QoS (Quality of Service) levels that will help operators grade services and give priority to premium customers who are willing to pay for a 'gold level' access to bandwidth and content.

11.5 Incumbent vs Greenfield

The incumbent operators will be trying to migrate their existing customer base to UMTS with minimal loss due to churn and battling various new threats and newcomers. The incumbents will have considerable strengths from knowing their customers and having established brands in the marketplace, etc. But the incumbent will also have to manage the thin line between convincing existing customers to move onto something better, without alienating them by having them feel that the operator has recently sold them something which was soon to be obsolete.

The competitive situation and proliferation of marketing messages relating to UMTS will become quite interesting with the emergence of the new entrants who are eager to take a strong dynamic and growing position in the marketplace. Most markets will see at least one new 'Greenfield' entrant and these have no legacy systems to integrate with and no legacy customer base which might become alienated. They start from a 'clean table' with nothing to lose and everything to gain. As they often have no established market awareness through an established brand, the Greenfield can take bold risks in establishing themselves. They will be taking an aggressive deployment strategy of both the UMTS network and services. And lets not forget an important fact; new entrants will be in subscriber acquisition mode and will be attempting to churn existing high value subscribers from the incumbent networks by offering higher value content and innovative service packages.

11.6 UMTS operator vs UMTS operator

While there will be varieties and niche players and particular quirks in any given market, the big picture will be the market share battle between the UMTS operators – incumbent operators, Greenfield operators, and MVNOs. The competition for market place will follow typical competitive rules from other industries. The mobile operators have not been strongly exposed to the rules of 'real' competition as the existing mobile operators and cellular carriers have mostly focused their full attention only on managing the dramatic growth in subscriber numbers and building up their network coverage. Few markets have seen full competition amongst four or more players of approximately equal resources – perhaps the UK is the best example of what mobile telecoms operator competition can be like, but even in the UK the battle between four rough equals is a relatively recent phenomenon.

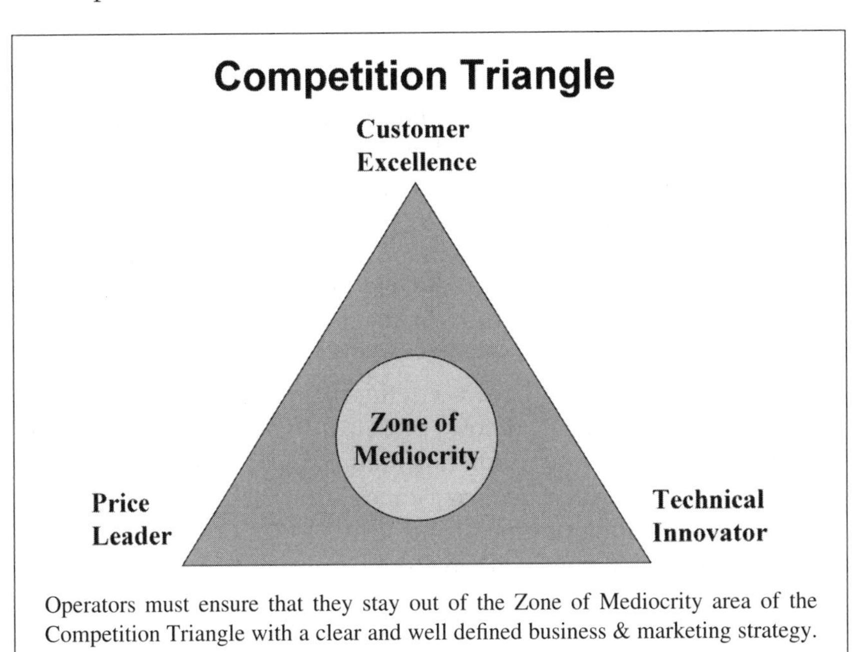

Operators must ensure that they stay out of the Zone of Mediocrity area of the Competition Triangle with a clear and well defined business & marketing strategy.

In UMTS competition most markets will have from three to six network operators and quite likely the markets will also see several MVNOs. Not all can be successful and not all can be winners when so many competitors fight for the same customer. A common theory of what happens in competitive situations is that there emerge three dimensions of perceived excellence. The marketplace assigns this perception of excellence based on how the players communicate and act in that marketplace. The three areas of excellence are price leadership, customer-orientation, and technical innovation. In most competitive situations of many players, one will emerge as the perceived leader on each of those three dimensions, and it is rare for one player to occupy the leadership position on two axis, and for any longer period of time it is impossible for one player to dominate on all three dimensions. A fourth area is called a 'zone of mediocrity' which is for players which are mediocre on all three dimensions but excel in none. This is the most perilous position to be in a competitive sense.

Price leader

A price leader needs to be active in the marketplace often promoting its price leadership position. The player does not need to be the absolute mathematical lowest price player, if it manages to convince the market of its price leadership in other ways, such as the timing of price reductions in lucky timing just before a major consumer survey of telecoms pricing levels, etc. The price leader will need to address the mass-market, meaning all segments in the consumer market but will also want to extend the price leadership into corporate customer segments.

The price leader has to have very efficient cost structures, from the infrastructure and technical costs of delivering calls and services, to the customer service and marketing staff costs of servicing its customers. Price leaders often take cost-innovation technologies into use, such as having billing on-line, using VoIP (Voice Over IP) technologies, etc. Price leadership is always a dangerous position because low price is easy to copy and a new competitor may emerge very rapidly. The price leadership player needs to be very sensitive to profits, and have very sophisticated profitability and price modelling tools to track the profitability and to prepare for competitive moves by the other players. Each market will have a perceived price leader but which player holds this position may change often. A price leader often falls into the Zone of Mediocrity and usually has little chance of returning to profitability at other dimensions except returning to Price Leadership.

Excellence in customer-orientation

The second area of market place success is for the player which is perceived to be best at understanding the customer. The excellence in customer-orientation is a marketing and sales-oriented organisation and this usually presents conflicts for those players who come to the game with very strong technical backgrounds. Most telecoms operators tend to be very technical in orientation and have a hard time adopting true customer-oriented and marketing-oriented processes. Initially this means that it is relatively 'easy' to be the best at customer-orientation in UMTS competition, as it is likely that most of the competition are not good at it. The biggest threats in UMTS to this position are likely to come from MVNOs – which are often pure marketing organisations – and Greenfield newcomer operators which often have headhunted some of the best talent from outside the traditional telecoms and thus often bring very strong marketing know-how without the technology oriented legacy mindsets.

It takes a lot of work with customers and expense with studies, surveys, and feedback mechanisms to build the market leadership in excellence in customer understanding. The player who has the strongest customer understanding is likely to have the lowest churn, which helps to off-set the costs of keeping abreast of what customers really want. The market position is often the most stable of the three positions of leadership, as it is very easy to spot when another player tries to move into a position of customer understanding and the established leader can usually easily increase its efforts to build up its lead. Once customer understanding leadership is established, it is usually a very profitable position, as the customers who appreciate better treatment are also less concerned about paying a bit more for the better service.

The player has to truly follow the edict that customer is king, and their customers do know it that this player treats its customers the very best. Customer understanding is of course something which every player claims to have, but can usually be easily spotted by looking at which player has the lowest churn rate. That is the player with the most satisfied customers. The customer-leadership player is usually the one who has most of the big or most profitable corporate customers. If it were to lose its position, this player is particularly suited to re-invent itself into either of the two other leadership positions, as it knows what the customers really appreciate, and thus could very easily focus on only key areas of price leadership or only on key areas of technical innovation and have a good chance to move to either of those positions.

Technological innovation leader

The third leadership position is that of the technological innovator. This player has to have a very modern and flexible technological platform and an innovative engineering staff. This player has to reward risks in-house and take chances by introducing innovative ideas to the marketplace more frequently than the other players. This is a favourite position for most technology companies, but only one can hold the leadership position at any one time, and that company is likely to be profitable. But the other contenders are spending a lot to try to take over the leadership position, so there is likely to be competition for who gets to be on top. The competition to get there may result in drastic spending errors in costly subsystems and pilot projects, etc.

The Technological Innovation leader often is the one which is a darling of the industry, wins awards, gets the best graduates lining up to join the company and has various very visible new product launches etc. Certainly when compared with the Price Leader and mostly also when compared with the leader in Customer Understanding, the Technological Innovation leader is the most ‘sexy’ player in the market and one most likely to be constantly in the news with exciting new services etc. Its market position is usually relatively stable over short periods of time and it can usually spot when a rival tries to invade its turf. But its leadership position is always inherently in danger of the introduction of more modern technology. This player cannot ever risk being cannibalised by innovation by competitors. It must be willing to sacrifice its current products of being cannibalised by its newer and better products. Depending on technical innovation overall, the technical leader can at times also achieve cost leadership. Whether the player wants to use that to fight head-on with the price leadership player or take the cost savings onto the bottom line and increase its profits, is a question for its strategic management.

Pick only one, not two

It is very typical of telecoms operators to say that all three axis are sustainable, and that they should pursue all three. While every operator of course has to be reasonably good at all of the three dimensions, in true competition, only one player can be the best at any one of the three. One is the price leader, one is the best at knowing customers, and one is the innovation leader. If an UMTS operator sets out as its strategy to take the leadership position on two of the three leadership positions, it sets itself out for three problems. First of all, it divides its internal efforts. Rather than focusing on one clear objective, the player will face two often conflicting objectives. The internal troubles caused by this type of ambiguity in objectives will hinder, sometimes paralyse entire departments. Secondly the player will face fierce competition from almost all other players. If the operator only selects one of the three dimensions, it is likely to face about a third or up to half of the competitors aiming for that leadership. But if the operator takes on two of the three, it will face most of the competitors who want one or the other (or some may also attempt both). Finally the operator cannot sustain a leadership position on two dimensions under competition for longer periods of time. Even if the operator manages initially to take two of the leadership positions, one of the competitors will soon win on one of the two – simply because that competitor can focus all of its efforts only on one dimension. This results in a loss of a stated position, bringing a loss of employee morale.

Zone of Mediocrity

The worst position in the marketplace is that of being in the Zone of Mediocrity. The player which is in the Zone of Mediocrity is no better than second best on any of the three dimensions, and often literally average on all three. This is the position where companies end up making losses. They have high churn rates because they are not leaders in customer understanding. They do not have the sales volumes that being the Price Leader bring. They do not have the market goodwill brought by technological innovation. They are the direct target of every one of the leaders, facing in effect a 'three-front war' and losing on all three. This player will suffer from bad employee morale, bad review from the industry analysts, and overall the position is one which makes profitable business almost impossible to sustain. If a player finds itself in this position, it has to select one of the three other strategies and pursue it vigorously to get out of this rut.

11.7 Making a start

Success in UMTS is not going to start when the first 3GPP (3rd Generation Partnership Program) commercial networks are launched in the second half of 2002 and in 2003. The foundations for success will have to be laid with GPRS services. Most industry observers feel that take up of GPRS is a prerequisite to strong take up of UMTS and that a slow start in GPRS will impact UMTS growth. This may be true but it will not delay the roll out of UMTS networks for a number of reasons.

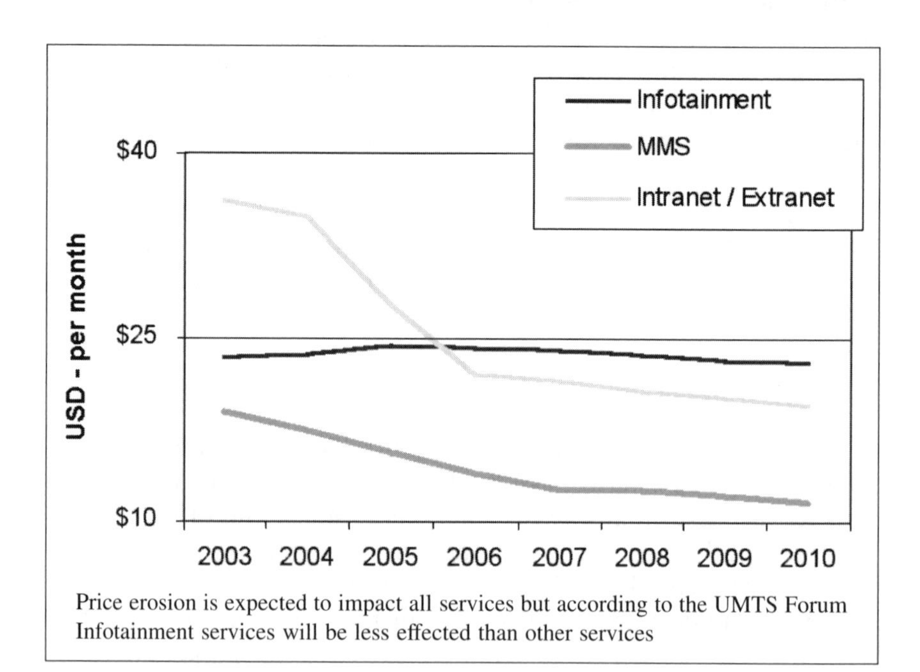

Price erosion is expected to impact all services but according to the UMTS Forum Infotainment services will be less effected than other services

A question of mindset

There is an old saying in telecoms, 'Sweat the assets and build the minutes'. In other words, run the system at its limits and get the customers to talk. This mindset comes from the old PTT (national Post, Telephone and Telegraph Company) days when fixed voice was a monopoly service and innovation was only used in engineering terms, if ever. This kind of thinking will kill the UMTS mobile data services business before it leaves the starting blocks. It takes a lot of courage for a senior telecoms executive who has been in the business for over 25 years to admit that their vision is too narrow and that preconceptions and legacy thinking could be a barrier to service innovation. Somewhere there has to be a balance between the reckless spirit of the Internet, the solid foundation of fixed voice and the evolution of mobile services. This requires real leadership at the top. It is a case of admitting that some of the younger executives in the organisation may have a better understanding of this new paradigm. That it is time to give them the ball and let them run, while still maintaining control of the direction of play and where the goals are.

Early operator examples

There are a number of mobile operators who have recognised that the rules are changing and that the old school of thinking no longer applies. NTT (Nippon Telephone and Telegraph) DoCoMo is without doubt one of these operators who is most often quoted and with good reason. By the end of October 2001 the number of i-mode subscribers in Japan had surpassed 28 million[1]. Of these over 7 million were iappli subscribers with Java enabled terminals. When we consider that i-mode was first introduced in February 1999 the growth is one of the most successful mobile data services in modern times. DoCoMo have created an environment where they control the access to content but they do not limit content. Just the opposite, they encourage new content and new i-mode sites and have a growing number of content providers who want to join the official NTT DoCoMo portal. The increasing amount of subscribers is attracting more content providers and the increasing

[1] The most up to date i-mode subscriber numbers can be found at http://www.nttdocomo.com/i/i_m_scr.html.

amount of content makes the service ever more appealing to subscribers, creating a virtuous circle.

There as some good lessons that can be learnt from Japan about how to create demand and to have a flexible approach to partnerships, revenue sharing agreements and pricing of services. However not all parts of the DoCoMo model can be used outside of Japan. There are cultural considerations to take into account like the fact that many Japanese travel for up to 3 hours to and from work every day meaning that they have time to browse various i-mode sites. There is also no messaging capability in the current PDC (Pacific Digital Communications) network and low PC penetration in the home. So access to e-mail via the i-mode mobile phone became the main driver for growth. Since then we have seen that entertainment services make up a major portion of revenue in the network. In Japan there is also a different disposable income criteria because cars and home ownership is limited by their relatively high price so people have more money to spend on other items like electronics and mobile phones. There is however some merit in the argument that if WAP had had colour screens and the 50,000+ content sites that i-mode have we would have seen better adoption and take-up.

The experience of Java based terminals has also been positive. So has the impact of i-mode and Java on ARPU. Figure 11.2 (NTT DoCoMo ARPU figures) show that the fall in ARPU has declined since the introduction of i-mode having a stabilising effect on revenue per subscribers. In Figure 11.3 we can see that the introduction of Java based terminals and the services that they enable have also had a positive impact on ARPU.

Although ARPU of the Java enabled subscribers has fallen since introduction at the beginning of 2001 it was still more than twice that of normal i-mode subscribers and we would expect that it will continue to have a positive impact on revenues.

More recently the introduction of the first 3G network by NTT DoCoMo, their FOMA (Freedom Of Mobile Access) service was launched officially on 1st October 2001, albeit with a 4 month delay due to technical challenges. Yet the news reports were that over 4000 terminals were sold on the first day of service. During the coming months we will witness how quickly the new services like video calls and video streaming are adopted. i-mode has given

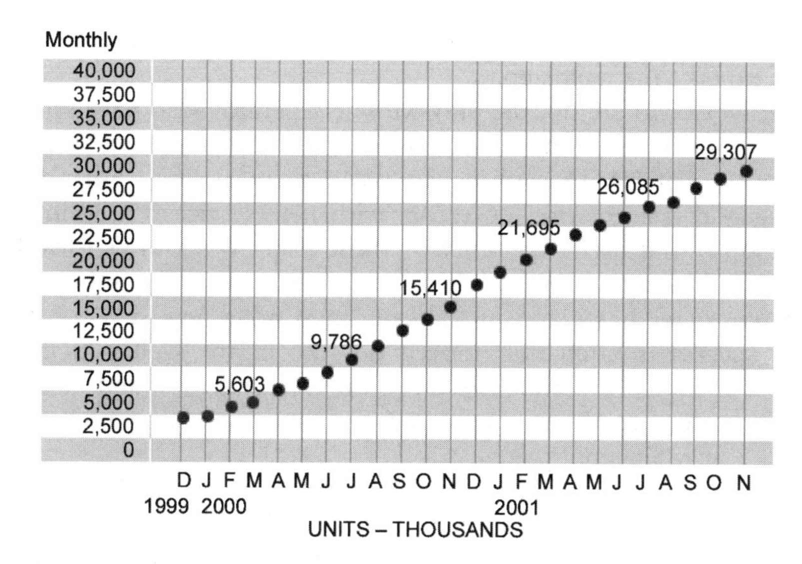

Figure 11.2 DoCoMO i-mode subscriber growth

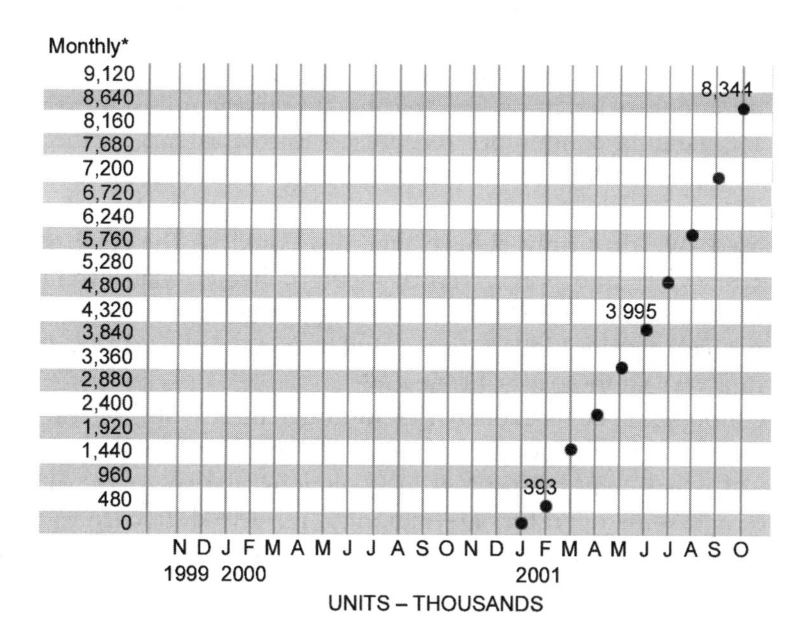

Figure 11.3 DoCoMo iAppli subscriber growth

the Japanese the taste of what UMTS can offer and now they are ready to take the next step.

Yet DoCoMo are not the only Mobile Operator who have understood that a new level of thinking is needed if they are going to make money out of UMTS services. In Japan there are two other very strong and fast-growing operators with new services on a similar philosophy and user experience as those using DoCoMo's i-Mode. KDDI and J-Phone have both been remarkably successful in building their customer bases and introducing innovative new services, but their success is often not noticed because of the well publicised success of DoCoMo. Similarly there are very successful and innovative operators in the Asian region including operators in Korea, Hong Kong, Singapore, etc.

There are operators like SMART in the Philippines who have introduced SMART Money, a mobile payment and banking service based on SMS[2]. As of the end of October 2001 there were over 500,000 SMART Money subscribers. SMART has been helped by the fact that there is a bank within their group and they have been able to avoid any regulatory issues. It is fair to say that in Europe at least the banking regulations are not making it easy for operators to offer or host financial services but we expect this to change. We think that this is only a temporary situation and that like the music industry, as soon that the banking community recognise the cost saving potential of mBanking and mWallet solutions and find a business model that works for them they will move quickly to adopt the solutions proposed. We do firmly believe that this could be the start of the cashless society and it is currently being delivered via current 2G technology. As 2.5 and UMTS technology enhances the user experience we feel that this kind of service will be used by all UMTS subscribers in a very short space of time.

We should maybe reflect here that although there is still a great deal of doom and gloom spouted about UMTS and the Mobile Internet, some of which may be justified, it is obvious that there are success stories and some operators are making a positive difference to their businesses with innovative and sometimes unorthodox approaches to the creation, delivery and consumption of mobile data services.

[2] http://www.smart.com.ph/.

11.8 Customer experience

The traditional customer experience in telecoms has been linear, proceeding very predictably from offering to buying to using and leaving the service. This is illustrated in Figure 11.4.

In the UMTS environment, the operator needs a totally new customer experience model. The operator does not only sell one service and provide it, but the operator sells and provides a wide range of new services, and does it all the time. The operator needs totally new models to manage customer relationship and processes. These are typically much more complex and intertwined, as can be shown in a simplified diagram in Figure 11.5.

The customer experience will become ever more important for success in the UMTS marketplace. Earlier on the main interest of mobile operators was to handle the unanticipated and quite dramatic growth in subscriber numbers, simply getting users connected, and building network coverage and capacity for them. In UMTS the situation will be very different. More competitors will be in the market, and the users are accustomed to levels of service, coverage, quality. The competition will shift to understanding the customer at various stages and having processes which address those needs appropriately.

On a general level one can categorise the customer experience into four main groupings Customer Understanding (i.e. Segmentation); Marketing programmes, (existing) Customer Intelligence; and Community Programmes. These all will impact the customer's experiences from seeing offers, to buying, to consuming and to leaving services. Table 11.1 shows these in table form.

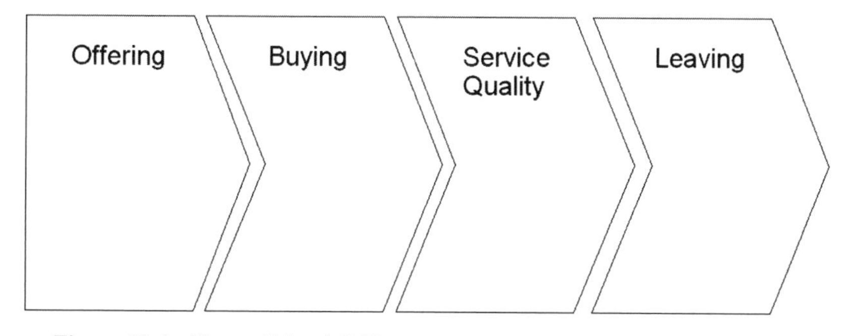

Figure 11.4 The traditional (2G) customer experience for a mobile operator.

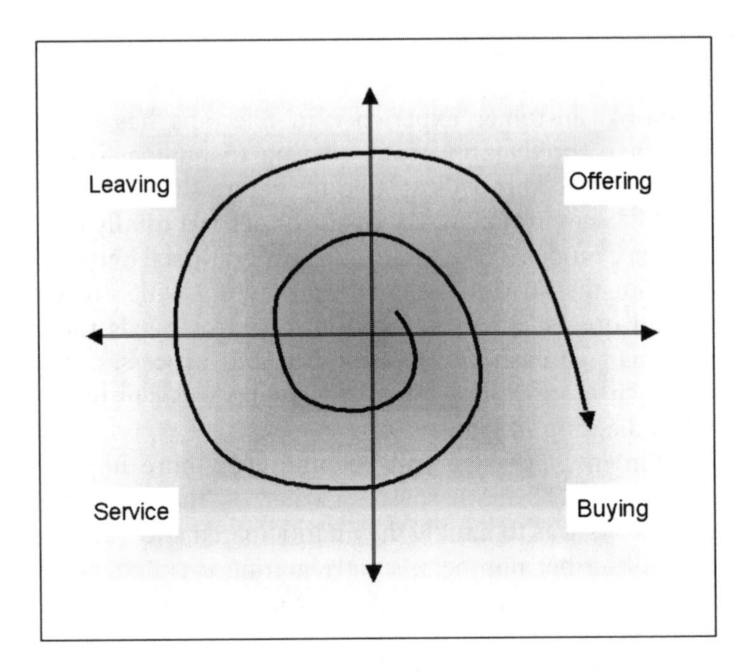

Figure 11.5 The customer experience for the UMTS operator.

As we can see in Table 11.1, the customer intelligence and personalisation have an important role in a successful customer experience. A customer must have an easy migration path to new services, easy to buy and activate, but also, for example, to transfer data from an old calendar to a new time management service. The operator must also know the customer to be able to offer new services and manage churn. For example, if a customer wants to leave a service it can be an opportunity sell a new more suitable service to him or her.

Table 11.1 Processes which are relevant in different phases of the customer experience

Offering	Buying	Service	Leaving
Segmentation	Distribution channels	Customer services	Churn management
Marketing programmes	Payment	Personalisation	Customer intelligence
Customer Intelligence	Activation	Payment/billing	De-activation
Community programmes	Migration	Quality management including 3rd parties	Migration

11.9 Handling churn

When mobile subscriber rates have grown at high speeds, it is perhaps understandable that operators have focused on customer acquisition. One should remember, however, that the cost to keep a current customer is a fraction of that to acquire a new one.

Churn, which increased in wireless networks by 4% last year – according to Cahners In-Stat Group – is the biggest problem facing the industry. In 2000 they predicted that 34 million wireless customers would churn as they struggle to find the best combination of cost, features and coverage. By 2005 that number could reach 80 million.

There is very little time to think of customers when senior management is in the boardroom hammering out deals – merger mania has played a major role in the lack of attention operators are giving to customer retention programmes. Perhaps in 2004 when churn rates reach a projected 41% globally (some operators are already at this level) the importance of retention will suddenly dawn on mobile operators and we'll see an emergence of good loyalty strategies and end users being rewarded for their loyalty. The operators who make this change in focus first have the best chance of gaining the most market share in the future. Some are already in this stage now.

By analysing real service usage and the segments which have a high churn, the operator can focus on those customers with a high propensity to leave. In this way the operator can understand reasons for the churn and develop its services and customer relationship to adjust to that knowledge. The operator must also understand which of the customers are important customers for a variety of reasons, and which of the customers amount mostly to costs. For example, a pre-paid user who does not use services might be insignificant, but an international businessman might be very valuable.

However it is not as simple as that. Figure 11.6 introduces what we call the Value Volume trade off.

This view can be taken for both customers and UMTS services. There are some customers and services that have low value but they make up for this in the fact that they consume or are consumed in large volume. There will be customer segments that are very low

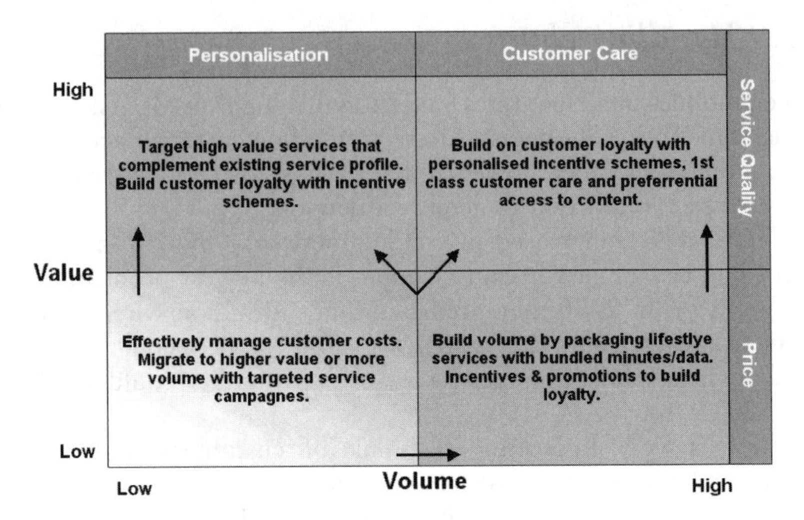

Figure 11.6 The Value Volume trade off.

usage and the cost of servicing these customers will be high. In some cases it may even cost more to service them than the revenue they create. The banks recognised this some years ago and in some countries introduced charges to cover this 'loss' that they were making in some areas. The customer anger at these charges has allowed many new banks to emerge who are offering better services to these dissatisfied customers. Likewise in the UMTS world operators have to fine ways to reduce their costs for high volume low value customers or service by having more efficient processes and logistics or by evolving the usage further up the value scale.

The strategy for these lower value customers is to provide minimum cost maintenance and to seek ways to increase customer usage by better understanding their consumption habits. Many of these customers may be in their early stages of adoption and will adapt their consumption over time. The secret is to understand the behavioural traits of these specific customer segments and find services that fit their changing needs.

The higher value (higher cost) services may not have the same volume when compared to others but can generate more revenue per se. Likewise the total volume of traffic may be lower in specific high value customers when compared to the total of low volume customers. In any respect it is necessary to recognise that there are different segments with varying amounts of value creating more or less overall volume in the network. Because of this growing complexity in the UMTS operator's customer base we would advocate that operators consider creating specific service evolution managers for each segment.

This kind of approach can focus the service creation along user segmentation needs so that the same service could be positioned differently to different groups. All this can lead to more understanding of customer needs.

So UMTS will allow operators to offer more advanced services with the potential to be personalised, localised and relevant to them at any moment in time. This will increase the emotional attachment to the particular operator/service package. In theory, this should reduce churn rates and deliver levels of loyalty that we have never seen before in the mobile industry.

Churn rates and forecasts

There is always intense discussion about churn levels. Figure 11.7 shows one indication of levels across different markets. The variation in churn rates can be due to factors such as handset subsidies; competition; retention strategies as well as different methods of calculating churn. These factors actually make it hard to compare like with like but churn does cost hundreds of millions of dollars per year for most operators. Suffice to say whichever way the numbers are analysed they make for frightening reading.

There are several factors which could lead to increased churn rates in the future which include number portability and increased competition. Number portability has been introduced in the UK, Netherlands and Hong Kong as of writing, removing one of the biggest barriers for the user to change operators. Increased competition results in pressure for operators which do not meet ever increasing demands by ever more sophisticated end users. These operators will have more competition than ever for their dissatisfied customers and would have to find ways to make customers profitable more quickly. In many cases some customers never pay back their cost of acquisition before they churn (Figure 11.7).

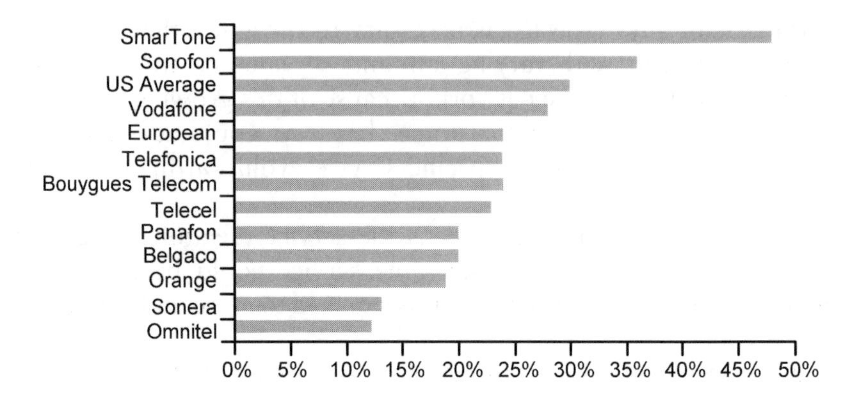

Figure 11.7 Churn Rates November 1999. Prodata.

Operators should not fear churn. Churn will happen and it is a phenomenon which needs to be understood and used. The prudent operator identifies potential churners, categorises them by who should be enticed to remain, who should be allowed to leave, and who are neutral. In a perfect world after a few generations of churn, all of the unprofitable customers have left and the most profitable customers remain. Of course that is an idealistic utopia, but at least operators and content providers should not fear churn, but learn to use it.

11.10 Winning on competitiveness

The UMTS environment is the most complex communication and computing environment ever created by man. It will allow the best of all existing technologies, richness of content, immediacy, relevancy and mobility. The convergence of the mobile world and fixed Internet brings the ability to bill and to have access to vast amounts of content. The migration from 2G to UMTS will bring the needed capacity to deliver the rich content and enhanced user experience. The convergence of content and delivery will bring the global media and content providers such as Hollywood, the games industry and newspapers into the UMTS environment.

The opportunities and the capacity provided initially by UMTS will totally overwhelm the ability of the UMTS operators to deliver a 'comprehensive' package of services. Initially there will be great differences in the offerings of UTMS operators within a country and also between operators in different countries. Eventually particularly successful and profitable service ideas will emerge. Early on it is very important for the UMTS operator and its partners to experiment and to follow what ideas take off in their own markets and in other markets. The UMTS operator should not try to be perfect, but rather be willing to experiment. When launching lots of new services rapidly, it is much better to find 7 out of 10, than 3 out of 3. The cut-and-paste thinking from the Internet, adopting and adapting will be the norm but there will need to be a keen understanding of how mobility aspects influence service adoption and evolution from the customers perspective. To borrow a thought from Marshall McLuhan: "A successful book cannot afford to be more than 10% new."

12

'A study of economics usually reveals that the best time to buy anything is last year.'
Marty Allen

Partnering in UMTS:
When You Cannot Do All Of It Alone

Frank Ereth, Jouko Ahvenainen and *Tomi T Ahonen*

The UMTS (Universal Mobile Telecommunications System) environment will have several operators deploying thousands of services in any given market. No UMTS operator can successfully manage all of the service creation needs by itself. Many feel that the UMTS operator cannot even handle all market segments successfully. As we have seen in this book, the UMTS environment is very different from that which current mobile operators are used to. Perhaps the biggest single change is that in the past the operator did everything themselves, in-house. In the UMTS area, that strategy is a sure-fire prescription for market failure. Only by becoming efficient at dealing with partners can an UMTS operator be successful.

As the operator will need to work with partners, this chapter looks at the various main partnership aspects of UMTS and its related revenue sharing options and examines how the operator can work with content providers, application developers, and MVNOs (Mobile Virtual Network Operators).

12.1 *New value chains and new roles*

New roles for the operator and its partners means new and more complex business models and revenue flows. The operator can generate revenue from several sources and each source will have different systems for handling related revenues and costs. A simplified view of the main different roles and their respective money flows is seen in Figure 12.1.

The illustration shows the separate positions of network provider and service provider which today are often the same UMTS network operator. The content providers will have access to the subscriber in many possible ways and similarly the subscriber will be able to access content through numerous portals. In some cases the content provider can deal with the customer directly. At other times the content provider goes through the service provider. And at other times the content goes through a content aggregator before passed onto the service provider. In each of these cases there are differences in who provides what value to whom, and then how the money should be split between the various players.

There is one view that the main role of the UMTS operator should be to create a high value business environment that draws in content providers and content aggregators. In some respects we agree with this role for the UMTS operator since without a clear and profitable

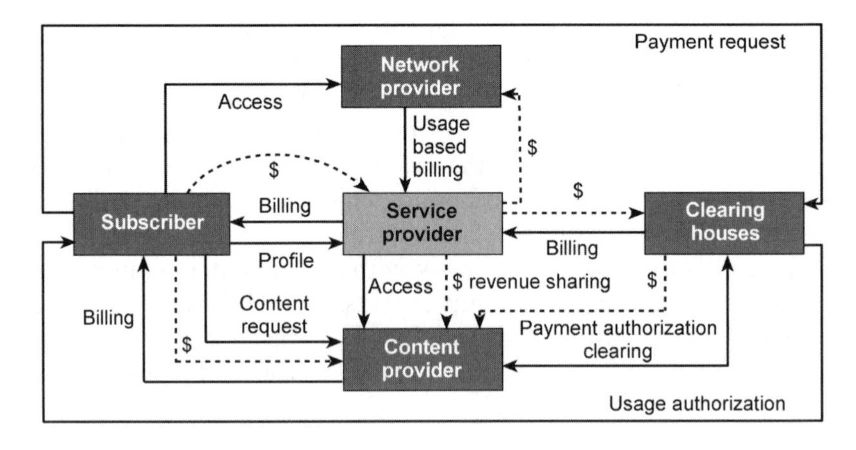

Figure 12.1 An example of revenue flows for the UMTS operator and service provider.

business opportunity for the content providers there will be less incentive for them to bring existing content to the UMTS environment and more importantly to create new mobile enabled content. The more attractive the UMTS content business, in other words, how much money can be made, the faster content will come to the market. The UMTS operator is the main gatekeeper in this area since they are mostly responsible for the revenue flow and how much of it is made available to content providers through revenue sharing agreements.

12.2 Partnerships

There are thousands of start-up companies with lots of eager young creative people with skills in mobile Internet programming. These people have good ideas on how to create new services, for example a mobile service to provide real-time event information and updates to event participants and event organisers. This type of service could be used by any kind of event organiser for any type of event, from the rock concert organiser to the ballet impresario to the professional sports club to the telecoms conference producer. The programmers who created the event information service would be able to help a mobile operator make money with the service. But to get the event organiser, the software maker and the network operator together requires partnerships. The mobile Internet in general and UMTS services in particular will be full of such partnerships, in fact *most* services will involve more than one enabling party.

The network operators, and to a lesser extent the major equipment manufacturers are not accustomed to dealing with partnerships, and even where they are, these tend to be with other operators or companies of similar size and style. Network operators are on a steep learning curve and the content providers and application developers, especially the smaller ones, are likely to be frustrated at many points early on while trying to form partnerships with the big operators and equipment manufacturers.

Partnering process

What is a partnership and what is the process of becoming a good partner of a big company such as a network operator or equipment manufacturer, in the mobile telecoms market?

The partnering process can be divided into the following sub-processes:

- Identify: defining the needs and finding possible candidates for the partnership
- Qualify: evaluation and qualification of candidate(s) for the partnership
- Integrate: establishment and combination of all the necessary business elements for the partnership
- Care: facilitation of the business functions needed to support the on-going partnership

Each partnership can be unique for the amount of partners, degree of partner involvement, extent of how long any partner will contribute to developing the common service, and how revenues and costs are shared. However for a successful partnership to work its basis must always be a win-win situation for all parties, where each partner gains out of the venture.

One example of how to create a realistic business case in what is becoming a more complicated value system is shown in Figure 12.2.

Impact on content partner

Typically a start-up company will have a good idea for some end-user service (event information, mobile translator, weather forecast, mapping software) and tries to approach the big telecom players (operators/carriers, equipment vendors, ISPs (Internet Service Providers)) and build the service or product. The content provider usually approaches with an idea of how the eventual solution could be used and sold, and have an initial business plan.

However, to create a good business plan for each of the partners in the UMTS environment is not that easy. The major part of the business plan is the business case, which contains the revenue and cost analysis for all parties, and the mere options are numerous with far-reaching impacts.

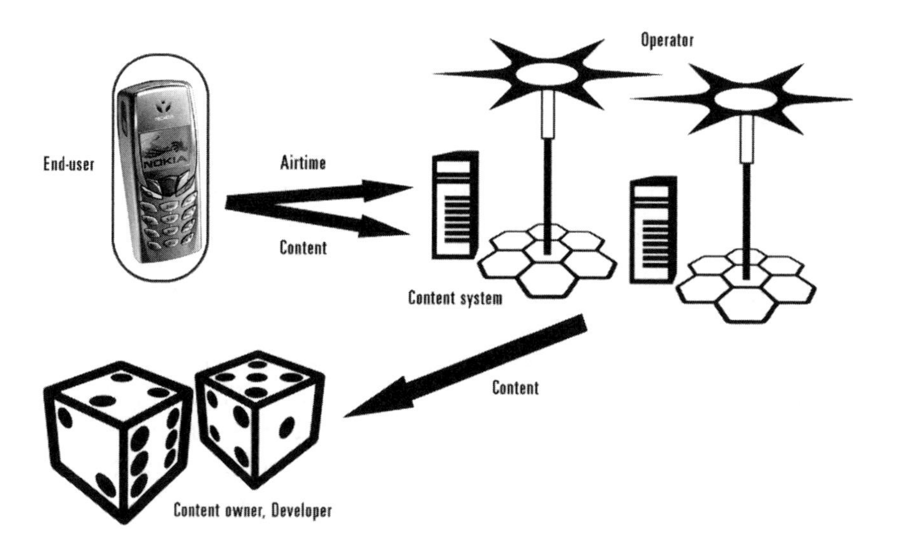

Figure 12.2 Typical service value system.

The revenue of the operator, content packager, systems integrator and content provider will depend not only on how much the end-user can afford to pay for each service but also how much value the end-user places on the service or content at any particular time. Each consumer customer has a limited amount of disposable income and each business customer has a limited budget for telecommunications spend. This has to be understood before any business proposal for revenue generation and more importantly any revenue sharing agreement is put into place. The challenge for the content provider who is coming from the Internet world is that the mobile world is quite different and end-user expectations are different.

As companies move more into the mobile service delivery business it becomes increasingly important to understand the complexities of the existing and potential mobile subscriber base and changing behaviour patterns especially in the youth market. As this market and subscriber knowledge grows the potential impact of any new service on revenue generation can be anticipated.

Impact on operator

The operator cannot produce the thousands of services that we will see delivered by UMTS networks by themselves. It could even be argued that some operators will struggle to provide many of the new UMTS services without a great deal of help. Operators are moving into an environment where they have to recognise where their core competencies lie, what competencies they can quickly grow organically or by recruiting new talent and what the competencies are that will come from third party partnerships. This can be a difficult task. Operators have had all the cake to themselves in what has been a growing and highly profitable market. Sharing ownership of this cake even in part to external parties requires senior executives to understand that the UMTS cake has the potential to be much larger, and will require new skills and competencies in order to make sure that their market share is in line with their goals and ambitions.

So operators need to find partners who can help them meet their strategic goals and ambitions for UMTS. The most profitable early services can be identified as those that are aligned with the 5 M's (Movement Moment Me Money Machines) as outlined in earlier chapters and exploit characteristics and deliver values which cannot be replicated with competing technologies, including the fixed Internet.

Operators and content providers need to evaluate the business opportunities in their individual markets. Joint market research or market estimates should be made continually to create a solid understanding of how each service could be delivered, and what price would the user be willing to pay for it. After this knowledge is secured, the partners should discuss revenue sharing mechanisms bearing in mind the value that each party brings to the table.

Network operators fear future of bit pipe

Those dealing with operators need to understand the fear that some network operators have about the future and becoming a 'bit pipe'. The operators recognise that one theoretical scenario is that the operator gets no revenues from the content, and only gets paid a set minimal transmission charge for mobile telecoms traffic. This bit pipe role is not appealing to operators who can see their opportunity to move 'up' the value chain and gain revenues from content. It is important to keep in mind, that most operators will approach any new service idea with the concept that the operator must gain something more than just the transmission of data.

The network operators' sometimes 'selfish' position on keeping revenue from content can lead to an overaggressive position. This may manifest itself as very dramatic requests in revenue sharing proposals where the operator might be reluctant to offer realistic splits of revenue with the content provider or application developer. The operator needs to recognise that an unreasonable initial position is counterproductive. To encourage third parties to bring new content and ideas to the mobile market there has to be a large business potential for third parties to tap into. In fact the larger this business potential the more this will encourage new investment in mobile content provision from a great many directions.

We do need to state one obvious point. An operators does not necessarily need to worry about becoming a 'bit pipe'. The bit pipe business can be very profitable – especially for one or two of the players in any given market – and can be combined with a focused service creation and revenue sharing strategy that builds on the core competencies of the organisation.

12.3 Revenue sharing

Revenue and cost sharing issues are at the core of any partnership. If the combined service generates many thousands or millions of dollars, how is this money divided between the UMTS operator and the many partners? Revenue sharing should be seen as a method of pricing services and products in a way that brings new value to customers and growth in services to all partners.

Revenue sharing is not simply a percentages of the total revenue shared, but it could also be based on:

- Monthly fee
- Yearly fee

- number of subscribers
- number of transactions
- number of sessions
- number of hits
- a one time fee after every sales
- part of a bundle

Each service may be support by one or a combination of the listed items. For example, a Chat service running in messaging platforms could be based mostly on an SMS (Short Message Service) tariffing method, while a City Guide Information service tariffing could be based on number of sessions, independently from the length of the sessions or from the size of sessions. Any revenue sharing agreement needs to identify what parts of the revenue are shared and those that are not. What will be counterproductive to growth is if UMTS operators are looking for up front payments for access to their mobile portal from content providers.

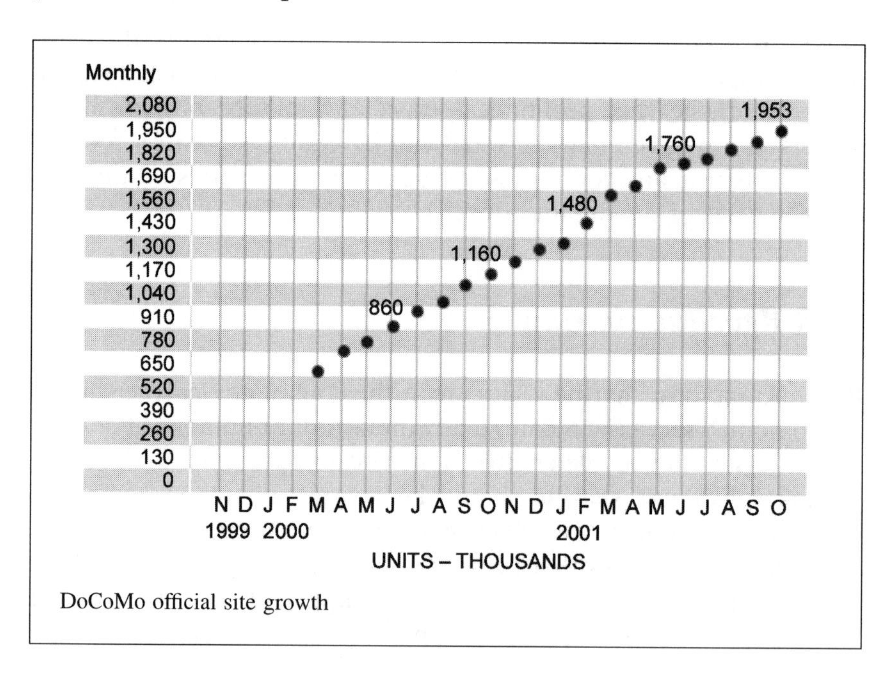

DoCoMo official site growth

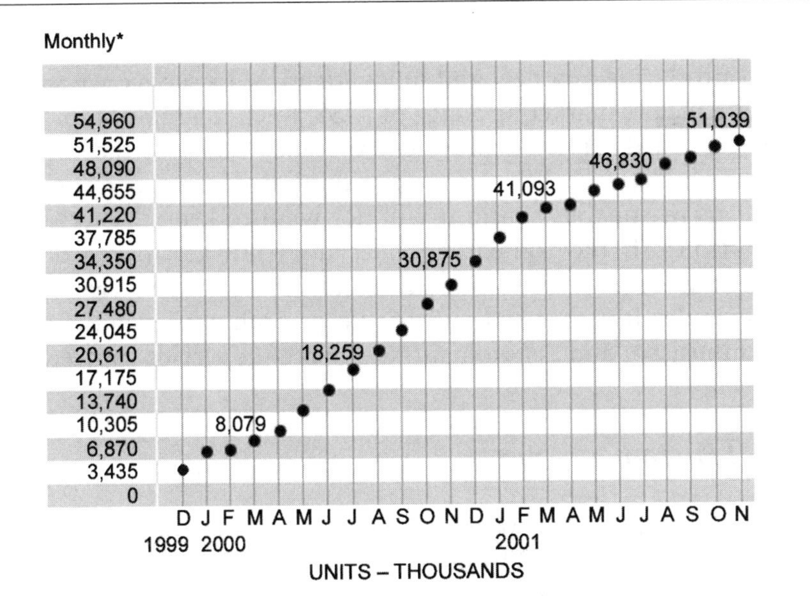

One of the success stories for DoCoMo has been the amount of content that they have had available from official and unofficial sites. As of the end of October 2001 there were almost 2,000 official sites on the i-mode portal where DoCoMo handle the billing (taking 9% commission) and over 50,000 unofficial sites.

Advertising/sponsorship adds another dimension

Advertising adds another layer to partnering and revenue sharing. Part of the revenue or even all might come from the sponsor, promoter or advertiser. Any advertising or promotion revenue charged to the advertising party is quite likely to be based on the number of end-users who receive the advert or promotion or even on the number who respond. Advertisers are always looking for ways to target their advertisements and promotions the response and see high value in the profile of each subscriber who takes up the 'offer'. This information can also form part of the revenue sharing agreement. For more on the advertising dimension, see the Money chapter earlier in this book.

As each new service is evaluated it is important to verify whether the billing and charging possibilities of the operator support the particular tariffing and bundling opportunities and if the service offer also fits to the buying profile of the target segment. It could be that the initial proposal for billing is too complex for the end customer to understand and this will impair the take up and growth of the service. Especially when introducing any new and innovative UMTS service we would advocate the old marketing adage of KISS. Keep It Simple Stupid.

For sustained service deployment the end results of the revenue negotiations have to be a win-win situation. If the operator does not get its reasonable return, it will not want to keep the service in the portfolio. If the content provider does not get its reasonable return, it will not want to maintain and develop the service. What is reasonable? It is that amount that keeps both parties in the game with enthusiasm and determination to make the service succeed.

12.4 *Trusted partner*

To further strengthen its position, the operator can act as a gateway towards third parties and be the partner for a wide range of contents which its users can access via its portal. The operator can add value by offering billing support for micro-payments, user profiling, provisioning services, authentication brokering, location-aware service delivery, statistics and a secure transaction thereby getting more volume in the network and more revenue. An example of this could be **community services** where there is a large number of small organisations like horticultural societies, sports clubs, activity centres and others too many to mention that have members who would like to be kept updated on events or news via their mobile terminal.

Likewise the operator can be the trusted party towards its users to ensure a smooth service by dedicating resources for third party management. Mobile subscribers are ideally thought to trust their mobile operator to handle commercial transactions and financial information as competently and confidentially as their banks do. Most importantly, the operator can offer their customers one stop shopping, single bill, anonymity and convenience.

Building a sound charging structure from the start

It is important to be able to tariff applications rather than only the bearer services. There will be new ways of paying for the content and services used by mobile users. The operator can charge per packet [of data], per event or by the amount of data [volume], by time of day and even according to the type of content or the value of a transaction.

There is a need for content and service specific charging – in addition to access charging – both in pre- and post-payment systems. Thus, the operator has to ensure that the service management and CCB (Customer Care and Billing) systems are able to meet the goals set by its new data services strategy. The operator needs to pinpoint at least these areas and take care to:

- Build sound charging, rating and billing systems from the start.
- Find solutions for prepaid services (of particular importance in countries with existing 2G prepaid systems).
- Have the rating system separate from the customer care and billing system.
- Be prepared to provision the service among multi-tariffing schemes (loyalty, sponsoring, revenue sharing, roaming, discount etc.).
- Establish user self-servicing from the start.
- Prepare a flexible charging and rating system for revenue sharing as services and players will grow in multitude and complexity.

Flexible platforms to facilitate service evolution

The operator should have control over service enabling and provisioning platforms so that services can be provisioned and personalised for the customers when requested. In the environment where there are many end-user applications and different rating schemes, it is important that the operator is capable of fast implementation, quick changes and flexible integration. Initially the flexibility will be less than ideal and services have to be tailored to meet the capabilities of the delivery, provisioning and billing solutions. As these systems quickly evolve more complex service evolution will happen.

12.5 MVNO (Mobile Virtual Network Operators)

An MVNO is an organisation that sells mobile subscriptions under its brand but without owning the physical mobile access network. However, MVNOs depend heavily on the willingness of physical

MNOs (Mobile Network Operators) to cooperate with them. As such MVNOs have to demonstrate that they have either a strong brand or an existing subscriber base that offers an opportunity for extension and increased revenue for the mobile operator.

The mobile infrastructure is more complicated to 'unbundle' than the fixed networks. The connection point between MVNOs and MNOs is dynamic, as the mobile user can appear everywhere in the network. Unless regulators do not impose MVNOs through telecoms regulations and enforce interconnect prices with the MNOs, the mobile network operators will always have the power to deny MVNO access or to impose commercial and contractual conditions.

The analogy for comparing network operators to MVNO's is akin to comparing airlines to wholesale ticket sellers. If an airline has excess capacity on a given route, it might want to sell a portion of its excess seats to various wholesale ticket sellers such as tourist agencies specialising in arranging group tours. In a similar way where network operators are likely to have considerable excess capacity early on in the lifespan of an UMTS network, a valid case can be built for selling some of that capacity for MVNO's to fill capacity and generate revenue. Of course much like the wholesale ticket agency with airlines, the MVNO would get to decide what it actually tries to sell, and at what prices, and how it markets itself. So the network operator will have to accept that the MVNO would sell to the same general market as the network operator, often at lower prices as is shown in Figure 12.3.

The current case of Virgin Mobile in the UK shows an exemplarily case of how the physical MNO can be actively involved in the MVNO set up. One2One was the 'fourth' MNO in the UK with a market share of around 17% at the end of 1999. Since then they have established, together with Virgin, the Virgin Mobile Group (50% ownership One2One, 50% Virgin) which launched its services in November 1999. By doing so One2One have made use of an existing brand proposition and have additionally profited from the distribution infrastructure of Virgin. It is interesting to note that many Virgin Mobile subscribers believe they have a higher quality of service when compared to One2One subscribers even though they are using the same 'physical' network. Virgin Mobile are now expanding into the US and Asia.

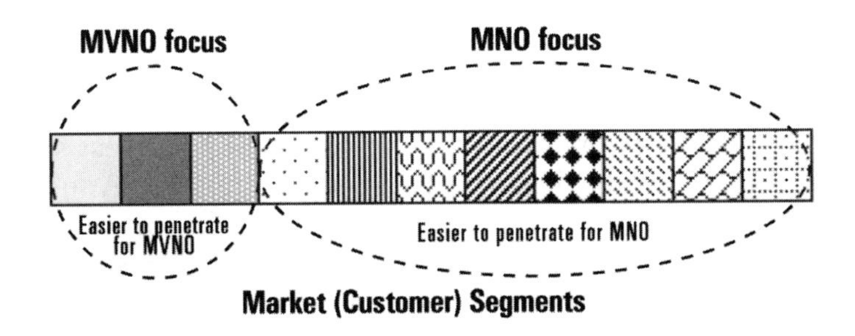

Figure 12.3 The value add of MVNOs for MNOs (Mobile Network Operators).

MVNO services for 2G and for UMTS

In current second generation networks (2G & 2.5G) a multiplication of new services are appearing which can be especially attractive to MVNOs. Some of the most interesting for the MVNO are end-user and location based mCommerce and mMarketing services.

As shown in the previous chapters, the UMTS opportunity will result in an explosion of new services. This gives a huge opportunity for any MVNO to attempt to enter the market. The MVNO could develop unique bundles of services addressing customer segments which may not have their needs met adequately by other service providers. This may be the case if the MVNO has strong customer contacts and in depth knowledge of its customer base.

A real example of a targeted niche MVNO from the UK is the MVNO set up by the Financial Times (known as FT). FT the newspaper and its website have a very high brand loyalty and could be viewed as one of the most successful independent Internet and Mobile Content sites. The FT MVNO is targeted at the readership of the FT and can be seen as a good example of a **finance services bundle** as the FT MVNO has as its core services those which its readers are interested in, stock information, business news, etc. The MVNO also provides other telecommunication services. These type of content providers are naturals for mobile services, especially relating to stock market and financial information and transactions. The MVNO is targeted only for the FT heavy user and FT has teamed up with mmO2 (previously BT

Cellnet) and the Carphone Warehouse to provide the niche market special phones and service.

Also, as was discussed in the marketing chapter, segmentation needs will increase dramatically for the network operator as segmentation will approach the theoretical ideal segment of one. As needs for segmentation continue to increase, some network operators might find it ever more challenging to satisfy all segments. Hence working with an MVNO set-up becomes a viable alternative. MVNOs have the possibility to create their own service portfolio for market segments that are maybe too small or too difficult to understand for MNOs.

Early MVNOs are likely to be major marketing companies which attempt to achieve a national footprint and national consumer market recognition via a strong brand and big marketing effort. Only time will tell whether such MVNOs will manage to enter and remain in the UMTS market place, and how many of such players will establish themselves as national and maybe even global players.

Remote metering services

For example with telematic services, a small player might develop a particularly strong service idea to address the needs of remote metering and tracking. Typically the service would involve small reader devices connected via the UMTS network. These could be fixed devices in difficult-to-reach locations such as weather tracking instruments, electricity and water usage meters, **parcel delivery and receipt devices**, etc. The UMTS devices would not need to have voice, nor colour screens. These devices could be very 'dumb' or depending on application, could be very intelligent but would typically be custom devices, not standard UMTS terminals.

The management and reporting of such data would require specialised usage, traffic handling, or delivery channel applications and software, and very specialised file servers and other IT technology. Similarly the maintenance and upgrade needs for this type of service could be very different from tracking for example the usage of a gaming application on the consumer services side. Finally the selling of this type of service would require a specialised sales force which would need to visit with each such customer, truly understand their business, and be able to argument the benefits of the new technology solution over the existing ways to collect and handle the data. So the sales of this kind of service could pose considerable stress onto a major nationwide national network operator. It is likely that a targeted approach by a specialist MVNO would serve the customer's needs best.

This kind of MVNO is not a full mobile service provider and in many respects could be seen as a value added re-seller into specific niche or vertical markets that are too narrow for the MNO to address economically. What the mobile operator gets is additional data volume into the UMTS network and providing this kind of partnership is profitable to both parties, and does not negatively impact the mobile operator's other more profitable services it can work out very well.

Will MVNOs succeed?

At this point in time it is difficult to assess how the markets will adapt MVNO offerings since mobile operators are keen not to lose too much control of their customers. Especially in well developed markets with penetration levels above 60–70%, churning subscribers from competitors is the only way to increase connections and business. Therefore some operators have opted for time limited contracts with MVNOs (e.g. Orange and MVNO Energis have agreed on a 3-year contract for the business sector in the UK).

In the end however it will be the end users that decide which of the players have the most appealing offering. And does the end user care who owns the network?

12.6 ASP – Application Service Provision

Over the next few years there will be a fundamental shift in the way individuals and processes within businesses access and use information. This shift introduces a major new market opportunity for making business in ASP (Application Service Provision). The mobile operator can take advantage of this emerging opportunity.

18 Vignettes from a 3G Future

Intimate Greetings from the Wife

I travel a lot and spend a lot of nights away from home. I ended up missing our wedding anniversary and sent flowers to my wife to mark the occasion. She surprised me by sending a personal intimate photograph of herself in a new night gown, to remind me of how she would be welcoming be back home. I love the fact that we can send very private and personal images to each other, and they arrive in practically real time.

An operator should recognise that the users will attempt to use the video parts of UMTS services early on to experiment. A far-sighted operator will enable low-cost video greetings and video calls by members of close communities, especially families and employees of any single corporate customers. Users will find many innovative uses for the camera and communication abilities of the UMTS phones and devices if they are not priced to be prohibitively expensive.

While the 'Internet revolution' and e-business has been talked about for several years, the whole economic impact of e-business is still to come. This market is enabled by the convergence of several key technologies such as UMTS, third generation mobile access, enhanced security and the availability of a ubiquitous and increasingly sophisticated Internet platform.

Driven by the need to achieve competitive levels of efficiency, businesses will strive to implement mobile data applications into many of their current processes. Time to market and skill availability issues will result in business applications being increasingly outsourced. By 2003 applications hosting will represent 22% of applications purchases. As well as high market growth, this represents a major discontinuity in several existing markets as they converge into one applications industry.

In addition to current applications such as CRM (Customer Resource Management), and ERP (Enterprise Resource Planning), many internal and external processes will become automated as part of new data applications. In addition to these new applications there will be a broadening of the base of businesses who use data applications, to include all types and size of business. Business users will expect to be able to use any device to connect securely to any application or information, from anywhere.

The application market is in its early stages, and most analysts agree that the potential is extremely large.

The usage of data applications will result in strong growth in the data connectivity market, as well as the promotion of the growth of a new Applications Service Provision market. Annual revenue from the hosted business applications market is forecasted to be approximately $10 Billion by 2003[1]. This represents a significant opportunity for network operators and service providers, as the major external influence on the adoption of wireless data applications is the network provider/Telco.

Today most business applications are tailored by system integrators for large businesses. These businesses increasingly provide access to data applications from the corporate Intranet. Corporate customers require access to applications from mobile devices, and opera-

[1] PMP Research, Mobile Commerce in Europe.

tors must thus provide secure corporate connectivity services. Smaller businesses, however do not have the resources to develop and implement their own applications. Nevertheless, market forces require smaller businesses to realise the same efficiency savings as their larger competitors. A new model, the ASP model, is thus emerging to allow smaller businesses to access applications on a pay as you go basis.

ASP as an service opportunity for UMTS operator

The ASP model has both detractors and advocates within the applications industry. Enterprises will increasingly find that they are forced to use an ever increasing array of linked business data applications in order to match the efficiencies realised by their competitors. Many companies cannot implement the applications they require in timescales that enable them to remain competitive. In addition small and medium sized companies may not have the technical expertise or management resource to implement complex application developments in-house, requiring IT, software and telecommunications sourcing decisions.

As these requirements are developing, and creating the demand for a new applications model, the enabling technologies are also being put into place. UMTS mobile connectivity offers delivery of large amounts of information to any location in a variety of formats from voice to sophisticated graphical presentations. This, together with widespread adoption of new types of terminal devices will enable whole new categories of applications to be developed. Such new types of mobile device, including phones, communicators, PDAs (Personal Digital Assistants) and handheld laptops will provide a wide choice of display capability, and intelligent management software will present the information in the appropriate format for the terminal in use. Secure Internet technology enables simple client software to access any required application, from one or more service providers. New middleware industry standards allows common authentication and validation to a complete set of applications, as well as information sharing between applications.

Operators who either develop ASP capability or form a strategic partnership with an ASP will be able to offer business customers a service with real added value. A prudent and far-sighted UMTS operator will keep its ASP opportunities in mind when considering the business opportunities of any given services.

12.7 Last part on partnering

The UMTS operator will need to adopt thinking which allows efficient handling of thousands of services. As services are developed, partnerships become an important factor, and even opportunities from working with MVNOs and VAS (Value Added Resellers) should be considered. Operators can gain considerable extra benefits from its platforms with which it deploys its services, by using those platforms and structures to offer ASP services to customers and partners.

Each player in the new UMTS marketplace will have its strengths and weaknesses. In that new environment, the UMTS operator has a strong position especially from its billing system, access to and the relationship with the end-user. In the end it comes down to the operator which is best at adapting to the new environment by being responsive to market needs and competitor moves. To quote Charles Darwin in the *Origin of Species*: "It is not the strongest of the species that survive, nor the most intelligent. But the one most responsive to change."

13

'For every economist, there is an equal and opposite economist.'

William A Sheridan

Business Case for UMTS:

Revenues, Costs and Profitability

Tomi T Ahonen and *Joe Barrett*

Will it be profitable to be in the UMTS (Universal Mobile Telecommunications System) business? The growth numbers are staggering. Nokia has predicted that the mobile telecoms market is expected to grow into the neighbourhood of a trillion dollars (1000 billion) annual market by 2006[1]. Whether that number is off by a hundred billion dollars, or off by a year, it is still astounding growth. Before the end of the decade, most of the world's mobile telecommunications will be UMTS. Yet in January of 2001 there were no UMTS services at all. The predictions call for phenomenal growth, unprecedented in mankind's history. At the risk of understating it, the market potential for UMTS is vast.

One should remember that our industry has an embarrassing history of underestimating growth in all forecasts in the mobile industry. About 10 years ago forecasts had 30% as the theoretical maximum penetration rate for mobile phones in the wildest optimis-

[1] According to Phillips Primary Research Study, '3G Wireless Market Expectations'.

tic scenarios. Five years ago, personalised ring tones, simple picture messages, changeable handset covers and even SMS (Short Message Service) were not considered significant services by any industry analysts and almost dismissed as revenue opportunities by most operators. The only sure thing we can say is that the trillion dollar market is probably wrong, by how much and in what direction will be the topic of heated analytical discussion for some time to come.

13.1 Market drivers for UMTS

The so-called fundamental curves about the near-term penetration and usage forecasts seem to be in agreement. Global telecoms usage will grow. Mobile handsets are increasingly becoming internet-enabled. Soon internet-enabled mobile terminal (handset) penetrations will exceed the PC (Personal Computer) penetrations, and not long after that, the mobile terminal will emerge as the predominant means to access the Internet, if for no other reason, than because for an ever increasing majority of users around the world who cannot afford their own PC, the internet-enabled mobile device will be their only access to the internet.

Within a few years, mobile access to new and existing Internet content will exceed fixed Internet usage. Most major analysts agree on these general trends, the question is only one of time; when will the mobile penetrations pass fixed phone and PC penetrations in given markets, and with some lag behind that, when will mobile access exceed fixed Internet access?

Need for speed

The rapid roll-out of UMTS networks due to factors such as coverage requirements, license requirements, license costs, and competitive pressures, will make fast deployment of infrastructure, and fast roll-out of services crucial for success in UMTS. The market will be made, but the fastest will get to pick the best niches, segments and markets. In most markets the UMTS licenses will be given to as many or more mobile operators than licenses awarded for networks using earlier technologies. So there will be more competition. As the industry has consolidated, there are more global and strong regional players who have market savvy and technical know-how to introduce competitive services.

The trends and drivers will help create an exceptionally dynamic, competitive and fast-paced marketplace for UMTS operators and their partners. One key to early success is good preparation. An understanding of the business case is critical in understanding what are significant factors, and to help in prioritising work.

While all these industry changes are taking place, mobile operators are in a key position to exploit new business opportunities brought about by a UMTS mobile world. The *key assets* of the mobile operator in this 3G business model are the micro-payment billing infrastructure, a large end-user base, an established mobile brand, the users' location information, established dealer channels and, naturally, the mobile network infrastructure itself.

13.2 Business case for UMTS

When it is simplified enough, business case analysis boils down to revenues minus costs equals profits (or losses). UMTS is no different at that level. But looking a bit deeper reveals that revenues, costs and the issue of profitability all have significant aspects that need to be discussed. Each of these areas differs from those under previous generations of mobile network business. Revenues can come from core business or from other sources. Costs include capital expenses, operating expenses and various implementation expenses but new to the equation in many countries are the license fees paid in license auctions, and while not new in itself, the prospect of network sharing has emerged as a viable cost controlling option. As UMTS services will inevitably involve partnerships in some form or another, complex revenue-sharing and profit-sharing mechanisms will be introduced, so the bottom line evaluation is not that easy either. This chapter will give a general but simplified overview to the UMTS operator business case.

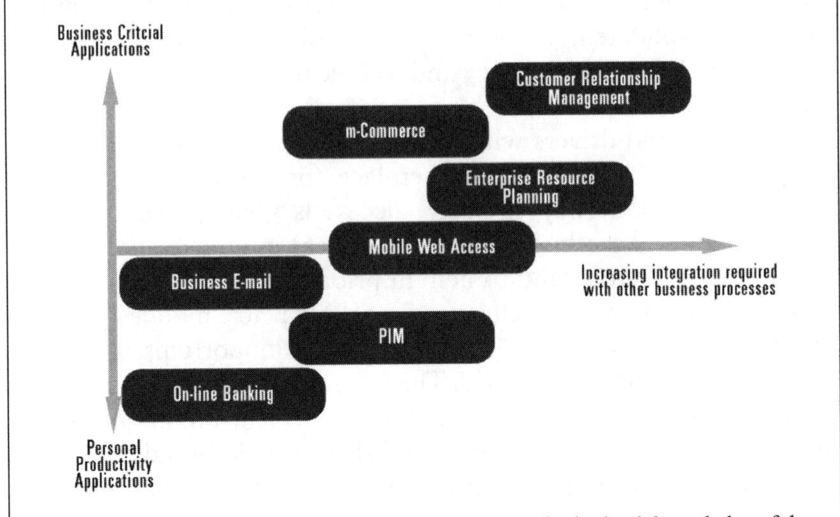

Mobile Business applications can be complex and require in depth knowledge of the corporate customer. This will probably be a specialised ''sell'' and may require dedicated sales staff with a strong knowledge of IT issues. This expertise is likely to come from good partnerships with system integrators.

13.3 *Services and revenues in UMTS*

Before discussing revenues it is important to look more closely at the types of services in UMTS. As the topic of this whole book is UMTS services, this paragraph will only discuss some general classes of services and how they behave in the network, to understand the different revenue streams that will emerge.

At the heart of the 3G experience will be the terminal and a new way of using the phone. The user will not just talk to it any more, he will be able to view multimedia images, watch video clips, listen to music, shop, book a restaurant table and even surf the internet. And, since it will always be connected to the network, there will be the instant delivery of important and timely information.

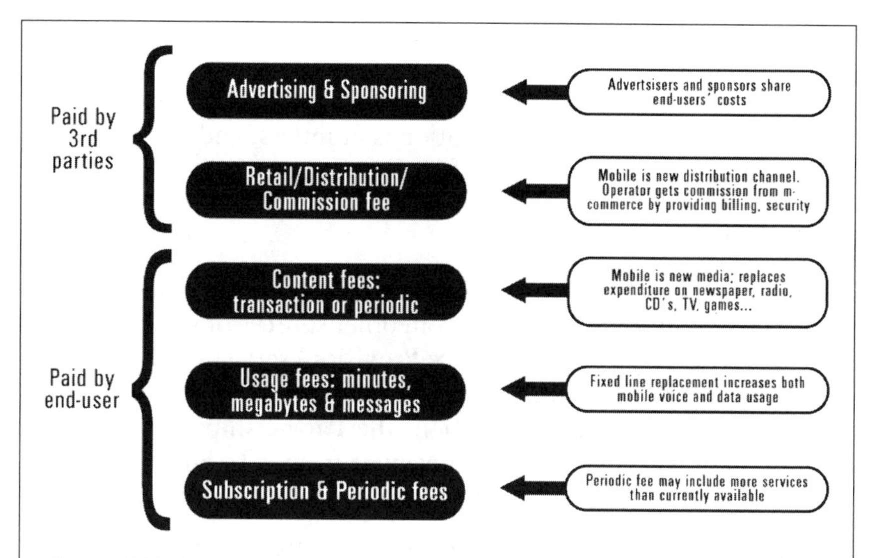

Source: UMTS Forum Report #9.
Not all revenue will come directly from the subscriber. There will be additional revenue from 3rd parties who will see the UMTS business as a new and exciting channel to their customers. In some cases 3rd party revenue will substitute or subsidies the UMTS service or will be in addition to the revenue also collected from the subscriber.

Service types by quality class

New 3G services are enabled with QoS (Quality of Service) classes for the terminal as defined in 3GPP (3rd Generation Partnership Project)[2] global standards. This model has several service classes ensuring that the radio connection is capable to support various types of applications:

- *conversational real time traffic*, such as multimedia conferencing
- *real time streaming traffic*, such as online audio/video reception
- *interactive traffic*, such as Internet browsing
- *background traffic*, such as downloading of mail

[2] 3GPP is the standardisation body for 3G technical specifications.

The reason why an operator will need to consider services by QoS types, is that QoS will have a direct impact on network traffic and potential for congestion. Note that for most service creation and marketing uses, the QoS classification is pointless, and categorisation such as discussed in the Types of UMTS chapter earlier, is needed.

Revenue types

Revenues in UMTS can come from core business – carrying traffic on the UMTS cellular network – or from other sources of income such as providing ASP (Application Service Provision) services and providing capacity for MVNO's (Mobile Virtual Network Operators). ASP's and MVNO's are discussed later in the Partnerships chapter. This business case will focus only on revenues from core business.

Core business revenues can be divided into several classes:

- Revenues in UMTS from services that already exist on current second generation cellular network systems. These would include services such as voice, and SMS text messaging. There are a couple of dozen such services but these will account for the biggest single portion of revenues for the operator.
- Services from newer and richer versions of current services, such as rich calls and picture messaging. Typical of these services is that user behaviour is likely to be similar to that on existing services, but the traffic load will be considerably greater. There will probably be under 100 such services.
- New services to current users. These include services such as voice streaming data services, and multi-user mobile games, etc. These services typically need partners to be deployed and will often have new types of revenue streams including advertising and sponsorship. There will be thousands of new services that fall into this category.
- New services to new users. New users with UMTS would mostly be services offered to non-human users, for example automobile telematics, or for machine-human interaction such as an alarm call system triggered by a machine alarm. The amount of users (devices) is likely to be enormous, but the amount of traffic generated by a single device may be trivial. There will be hundreds of services in this category.

- Revenues from mCommerce – paid by merchants as commissions to the operator. There will be several types of mCommerce transaction types, from ticketing to catalogue browsing to mBanking, etc. If counted by eventual mCommerce merchants, there will be tens of thousands of differing services. If counted by types of transactions, there will still be hundreds, possibly thousands, albeit overlapping with services listed above.
- Revenues from advertising and sponsorship – paid by advertisers to the operator. There will be many types of advertisements, banners, coupons, promotions, and sponsorship types. Many of the mCommerce and new service types will include some variety of advertising or sponsorship, so there will be hundreds, possibly thousands of varieties, overlapping with potentially all of the above.

Each of the above has different traffic patterns in how they generate traffic into the network, and each grouping has different factors in determining how much revenues can be generated. Some of the main assumptions are discussed here.

Assumptions about how traffic and revenues develop over time

According to the above general groupings, the following assumptions can be made about how UMTS services will evolve both in the amount of traffic and revenues:

Traffic for existing services will grow for the foreseeable future. As mobile terminal penetrations keep growing, and the substitution from fixed to mobile takes place, as well as the perception on the price of mobile calls keeps dropping, the amount of traffic will keep growing for services like voice and SMS. These will continue to be a major part of UMTS operator revenues for a long time. But price erosion from increased competition will drive per-minute and per-message tariffs down. Most views of the future suggest that tariff erosion is faster than traffic growth, which results in a gradual drop in the overall revenues from the existing services.

Traffic for new versions of existing services, such as rich calls and picture messaging, will grow initially at remarkable speed as has been seen with SMS and i-mode initial usage patterns. The speed at which new services grow will not so much depend on the technical aspects of the networks or terminals or the availability of new services but on the mindset of the operator's organization. Some operators will take a conservative view and base their developments from existing services. All UMTS operators will offer these evolved services and significant differentiation will be unlikely. Initial opportunities exist for higher prices, but competition will bring price erosion to these services quite quickly. Other operators will be able to throw off the chains of old PTT (Post, Telephone and Telegraph) thinking and push the boundaries of UMTS service creation.

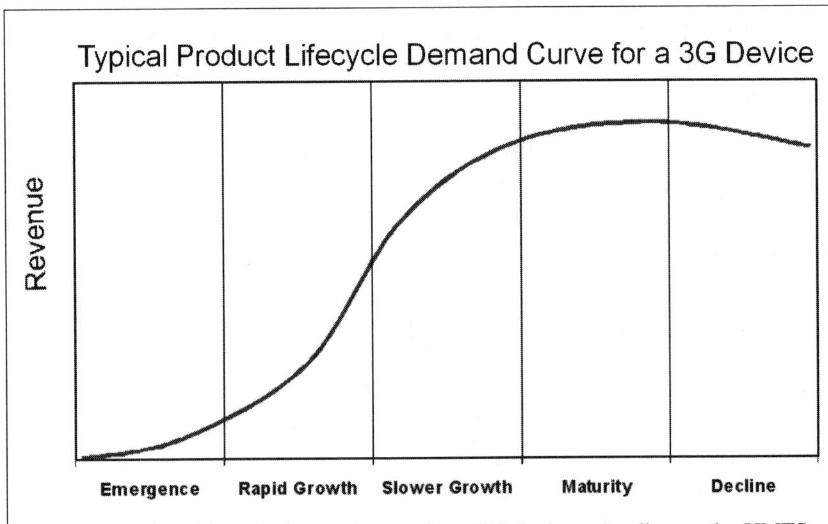

Typical product lifecycle demand curve for a 3G device according to the UMTS Forum report Number 9.

New services such as various information and entertainment services will emerge. These will vary more between the innovative operators. As the content providers such as major news-media, the music and games industry, etc., become familiar with the technology, they will soon push content that will fill the capacity for an individual session, so the amount of traffic will soon grow. Music streaming and video clips will become commonplace. Creative uses of built-in cameras will also produce a lot of traffic, as services such as 'Show Me' will appear. As newer and richer content arrives, so will the opportunities to generate revenues out of them. With new services, price erosion will not hit as hard as with established services.

Non-human services will be created for the machine population which will also generate traffic into the UMTS network. The machine population will keep on growing as advances in robotics and automation will learn to take advantage of mobility, the location information, and the video and audio capabilities of the UMTS network and terminals. Non-human traffic will often be small amounts of data, but there may be very many such tiny data transactions. Service billing for machines may be based on monthly fees rather than per-usage fees, depending on application. The machines rarely need immediate connections, and can tolerate delays of seconds or minutes, even hours in transmissions, so background classes can often be used in transmitting data. Therefore, non-human traffic, although large in data volume, will be relatively modest in revenue generation.

mCommerce is one of the areas of the UMTS business case which has the greatest variety in opinions. Some say it will be marginal, limited to ticketing and few very niche markets. Others say the mobile terminal will soon replace credit cards, and due to the micro-payment features, will actually deliver on the promise of the cashless society – something that the credit card industry was unable to deliver precisely because they did not want to handle minimum purchases valued in cents rather than dollars. mCommerce's impact to the business case should be counted only on the commission revenues that the UMTS operator would get rather than the full value of the items purchased, such as books, tickets, etc.

Advertising and promotion will generate additional revenues to the UMTS operator. The mobile phone will be the most personal promotion media ever, since it is the only communication device that is carried by the user 24 h a day. The advertisers could use profiling information with location based knowledge to provide targeted promotions that would be under the control and acceptance of the subscriber. Advertising and promotion by itself would not generate large amounts of traffic, but would be embedded or added to the traffic relating to games, film clips, music, information content, etc. Using revenue models from newspapers, the fixed Internet, radio, TV and games, it is reasonable to assume that similar patterns will emerge as in those industries, where advertising revenue will become a significant part of total revenues. Again as with mCommerce, opinions vary about the degree. One should remember that sponsored sites and sponsored free content models already exist on mobile services, and these are likely to become more popular over time.

13.4 ARPU (Average Revenue Per User)

One of the industry standard measures of revenues is the ARPU or Average Revenue Per User. A very commonly referred to ARPU graph from the UMTS Forum has been illustrated here to show how this business case ARPU evolves over time. It is based on most of the assumptions discussed above, but for more information on that analysis, please read the UMTS Forum Report number 9.

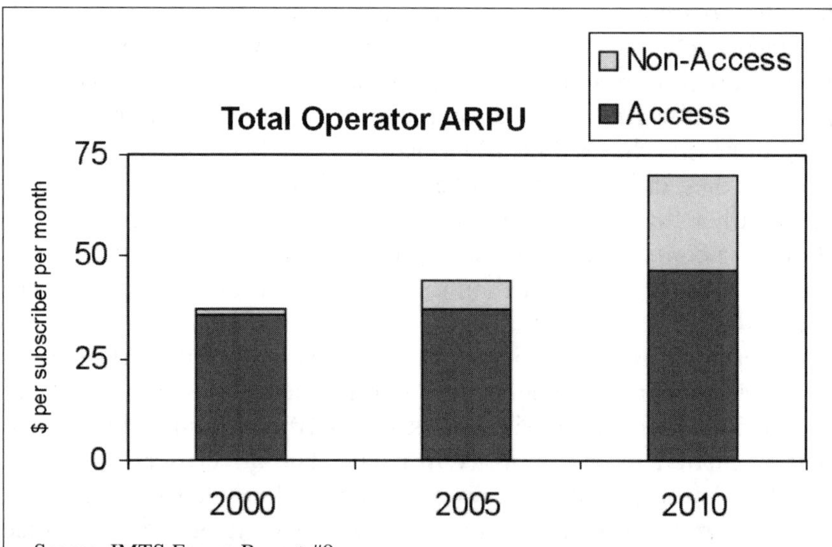

Source: IMTS Forum Report #9.
Average Revenue Per User (ARPU) has decreased in the past 3-5 years driven mainly by the increase in low usage prepaid subscriptions but ARPU now appears to be levelling off in many markets. Operators with the right strategy can look forward to increasing ARPU levels.

While ARPU grows, the consumer end-user phone bill will not grow at the same rate

It is easy to leap to the conclusion that when ARPU doubles, the end-user phone bill for the consumer – for you and me – would also double. This is not the case, however. First one must remember that the ARPU measures average revenues for all users. That includes typically a mix of business and residential users. Some of the services that are being introduced in UMTS will serve particularly the business segment, such as video conferencing and fast downloads from the corporate intranet, etc. These premium services would command premium tariffs, and generate more than average revenues. The consumer would typically not consume these premium services, at least not for any significant degree. That means that the consumer end of the total picture would have on balance less of the most expensive services, and thus less of the overall ARPU.

The end user phone bill for *current mobile services* – mobile phone calls and SMS services – will actually diminish. The overall consumer phone bill will probably grow modestly, but the growth will all be from new services. The new services come in forms such as richer content – if the message sent is no longer just text as an SMS, but includes a colour picture, then the user understands that it costs more.

The new services will come in forms such as substitution from spending that money in other media. For example if regular postcard and greeting cards are substituted by sending picture postcards, then that spending is transferred from buying the cards and postage stamps, to the equivalent spending on the phone bill. And there will be considerable new content items such as downloading music, for which the consumer is willing to pay.

These kinds of spending changes have already happened in some advanced markets especially within the youth segment. Teenagers in mature mobile markets in Western Europe are spending over 50% of their disposable income on their mobile phone bill. This means that they spend less on other items like clothes, going to the cinema and even on cigarettes and alcohol. In Japan one of the largest department stores, Daiei, has publicly stated that it has seen a noticeable drop in the spending of young people, directly attributable to spending shift to mobile spending.[3] These spending habits will continue to favour the mobile phone as a means of communication and transaction as they move into the workforce where they are earning their own money. These current and future spending habits need to be factored into the UMTS business case.

The operator will get additional revenues from many completely new sources, which are not billed to the mobile phone user. For example telematics traffic relating to commercial metering and instruments would produce traffic into the network, and revenues to the operator, but this new money would not come from the pockets of the mobile phone user, at least not directly. Similarly the operator would get significant new revenue sources from mCommerce transactions and advertising and sponsorship revenues.

So while operator ARPU follows an S-curve path, and grows significantly over a 10 year span, the consumer phone bill will grow only modestly, and for that growth the consumer will get considerable additional value and utility, which in many cases the consumer will experience as a reduction in overall spending. It is just that a higher proportion of the total household spending will go via the phone bill.

13.5 Costs

Costs for UMTS can be divided into five main types of costs. They are the costs of purchasing the network; the cost of the UMTS license; the costs of implementing; the costs of operating the service; and profit sharing costs of service partnerships.

CAPEX (CAPital EXpenses)

The CAPEX (CAPital EXpense) costs consist of the RAN (Radio Access Network), the core network, various IT (Information Technology) systems, billing systems, etc.

Two thirds of UMTS capital expenditure will typically be in the RAN so for the costs of building a network, this is the biggest area in CAPEX and one which has the biggest impact to the overall business case. The move to IP (Internet Protocol) based RAN in the network is

[3] Wired Magazine September 2001.

expected to bring savings to this part of the network. Coverage and capacity extension techniques and the early deployment of UMTS enabled equipment bring additional savings in CAPEX.

License fees

The license fees vary greatly from one country to another. In some countries the license fees have been trivial and the license has been awarded on merits in so-called beauty contests. In other countries there have been auctions which have produced auction bids for licenses that have grossly exceeded any expectations. In countries where there is a licensing auction, there have been cases where the license cost has been roughly as high as the cost of the total network infrastructure investment, effectively doubling the CAPEX. The license cost is the single cost element with the biggest variance and no clear pattern can be given to estimate the license costs. This information is public information and can be easily obtained from the regulators in any given country.

IMPEX (IMPlementation EXpenses)

The costs of implementing the network include various costs of installations etc. The biggest single component in implementation is the site acquisition cost. Site acquisition costs can easily equal the costs of all equipment to be used at the site, especially in downtown locations of densely populated cities. Many ideas have emerged to take advantage of co-location of equipment and even network sharing technologies, to reduce the costs associated with sites.

OPEX (OPerating EXpenses)

The costs of operating (OPEX) the network include all the personnel costs and other running expenses. The biggest factor in OPEX is marketing expenses. In countries where handset subsidies are a significant factor, this tends to be the biggest single marketing expense. Modern digital networking equipment is growing in complexity so a well thought out and well managed network is a pre-requisite for OPEX cost savings.

Overall, the marketing expenses in mobile telecoms have been in the range of 25–40% depending on market and the level of competition. Predictions of UMTS competition and OPEX tend to have the marketing expenses at roughly the same level. A new cost emerging with a growing share in OPEX is that of support IT technology to enable the new services. The links between the infrastructure teams and the service and marketing teams has to be strong. As the networks evolve towards IP standards the development of new and innovative IP-based services will create pressure to hire more IT staff to develop and maintain the various servers and platforms needed for operating the UMTS network. Again the linkages between all these teams need to be kept intact so that someone has a helicopter view of what is actually happening across all boundaries of the operator's business.

Partnership costs

The costs of partnerships include various revenue sharing schemes and profit sharing plans that are implemented with various content, applications, and other service providers who join with the UMTS operator in deploying the services. They need to be considered on an individual case-by-case basis for all significant services where an outside content provider or partner is used. Here the early trends are only emerging from WAP (Wireless Application Protocol), i-mode and early GPRS service development, so only general assumptions can be made about how revenues could be split.

13.6 Sensitivity analysis

Sensitivity analysis is used to isolate factors which have a large impact on the overall business case. The consensus in the industry is that the biggest single factor in the business case is the development of the revenue streams of the services, and thus ARPU. A small change in the ARPU will result in a dramatic change in the overall payback periods, net present values, internal rates of return, etc. It is vital for the UMTS operator to understand the services and how to nurture the traffic growths, and understand the value chains, and ensure the profitability of any given service. It is critical that the UMTS operator is early in

19 Vignettes from a 3G Future

Our Competitor News Update Direct to my Phone

The competitor analysis group at headquarters has again introduced a wonderful innovation for us to keep abreast of what is happening with our competitors. They have now produced a UMTS version of our Intranet-based competitor news, and it now sends a daily message to all who have signed up for it, with the headlines of the major competitor news of the day. I only receive the short headlines, but I can click on any of them to get the one-paragraph story. I can skip the competitor news of course, and I can also scroll back to find the previous news or search the newsbank. And now I am never surprised by my customers when they ask me about what do I think about what our competitor did today...

The various news organizations and publications can capitalize on UMTS of course, but an early adaptation is likely to be the urgent corporate internal news updates such as the competitor news, etc. For companies where the data is gathered and published on the Intranet anyway, and where the target audience is very mobile - such as executives and sales representatives - a UMTS version is the natural progression in the service. Those UMTS operators with large amounts of business customers might want to prepare packages and example services for their customers to replicate in their organizations. This will help migrate the executives to UMTS terminals early on, and help build stickiness and brand-loyalty to that UMTS operator.

any target service area, and that the operator is responsive to changes in the market.

For Operators to achieve a fast time to UMTS profits they will have to have synchronization of the networks, applications and terminals to deliver unique end user value. Speed is going to be the differentiation in UMTS. Not Mbits/s but speed of network delivery, speed of roll out, speed of applications delivery, speed of invoiceability and speed of responses to market changes.

Naturally, each business case is different and this is why operator have to use sophisticated modelling tools updated with the latest available data to run scenarios and plan contingencies. The key to success in operating a UMTS wireless network is to understand what content, what applications to develop and with what business models.

13.7 The future of ARPU

When mobile operators had only a few services and most users had one subscription or the equivalent, then the concept to the ARPU was a good tool to use in comparing operators of different sizes, in different countries, even using different technologies. But recent trends have made the ARPU thinking somewhat obsolete. Trends have emerged where users are not limited to only one mobile phone and only one subscription. Users often have multiple subscriptions for example one from work and another one for personal use. In some cases there is a primary phone and subscription for price needs, and another one for coverage, service or convenience needs. Some people call out on one phone and receive calls on another one. There is also the emergence of non-human users such as automobile telematics, home appliances, remote metering and robotics. The UMTS Forum Report Number 9[4] brought this issue to the forefront of the UMTS operator community. A common opinion is developing in the industry that ARPU alone is no longer accurate enough to be used as the only measure of operator revenues.

[4] http://www.umts-forum.org/.

Multiple phones means variety in behaviour

Early estimates have human use approaching 120% personal phone penetration. As part of the human population will not be having a phone – babies for example, typically estimates have approximately 80% of the human population owning phones, and roughly half of those owning two phones. That is how 120% is achieved.

The primary subscription behaviour may approximate that which can be measured with the traditional ARPU, but the additional subscriptions of phones used only in special cases, would yield very different ARPU patterns. If we calculate only the primary subscription, we underestimates the traffic generated by that user. If we calculates the secondary subscriptions, then we are strongly misled by what seems like a low-use new user. The best case would be to measure the total traffic of all subscriptions by that user. Or else the operator would need to survey and estimate by traffic which subscriptions are second subscriptions, and measure and compare them only to other second subscriptions.

Machine subscriptions behave very differently

When taking early estimates of machine SIM (Subscriber Identity Module) cards, the likely ratio of UMTS machines to UMTS humans is going to be of the magnitude of 2.5:1. This means that there will be over two times as many UMTS connected machines than there will be connected humans. But it is unlikely the machines will request to download any picture clips, movies or new ringing tones. The behaviour patterns of these machines will be quite different from human traffic patterns. Some can be predictable as a clock, others can be intensely productive and generate huge amounts of traffic for short times. But they tend to be different from humans.

While ARPU will continue to be one general measure of operator revenues, total revenues and segmented revenues will become more relevant. Perhaps an ARPHU (Average Revenue Per Human User) and ARPMU (Average Revenue Per Machine User) will be more relevant. Even here the ARPHU figure would need to be carefully constructed with correcting factors to account for Human User primary subscriptions and Human User secondary subscriptions.

Another option is to segment the ARPU per lifestyle segment or age group or a combination of the two. This can give a more diverse view of the ARPU demographics and a better view of where to focus the ARPU growth efforts.

13.8 Back to business

This chapter has examined the business case for UMTS. The case depends on very many variables and with the complexities involved a whole book could be written just on the business case for any given operator. In fact such books *should* be written at those operators in the forms of strategic business plans, network designs, marketing plans, etc. But for the general public and as an overview to anybody involved in UMTS, this chapter has attempted to show the importance's of services and revenues, what is the ARPU and how it can be used, as well as how it is likely to evolve as a measure. This chapter has also looked at the CAPEX and OPEX involved in the UMTS business. Before one can be successful in UMTS, one has to understand its nature. As the polar explorer Roald Amundsen said: "Victory awaits those who have everything in order: people call this luck."

14

'It should be possible to explain the laws of physics to a barmaid.'

Albert Einstein

Technical Primer to UMTS:
WCDMA Technology for the Layman

Harri Holma and *Antti Toskala*

This chapter is intended to cover briefly the main technical aspects of UMTS (Universal Mobile Telecommunications System), specifically from the WCDMA (Wideband Code Division Multiple Access) technology point of view. As the UMTS radio interface, the WCDMA technology is going to be the most widely used technology for IMT-2000.

14.1 Air Interfaces for UMTS

Within ITU (International Telecommunications Union), several different air interfaces are defined for third generation systems, based on either CDMA (Code Division Multiple Access) or TDMA (Time Division Multiple Access) technology. WCDMA is the main third generation air interface in the world and will be deployed in Europe and Asia, including Japan and Korea, in the same frequency band, around 2 GHz. The wide deployment of WCDMA will create large markets for manufacturers and the providers of content and applications. Note that in second generation systems different air interfaces are used in Europe, in Japan and in Korea: GSM (Global System for Mobile communications) in Europe, PDC (Personal Digital Cellular) in Japan and IS-95 in Korea.

In North America the spectrum around 2 GHz has already been auctioned for operators using second generation systems, and no new spectrum is available for the new systems. Thus, third generation services there must be implemented within the existing bands by replacing part of the spectrum with third generation systems. This approach is referred to as re-farming.

In addition to WCDMA, the other air interfaces that can be used to provide third generation services are EDGE and multi-carrier CDMA (cdma2000). These solutions are targeted mainly for providing third generation services within existing frequency bands, for example, in North America. EDGE (Enhanced Data rates for GSM Evolution) can provide third generation services with bit rates up to about 400 kbps within a GSM carrier spacing of 200 kHz. EDGE can be flexibly used together with GSM in all existing GSM bands, like GSM900 and GSM1800. EDGE is expected to be used especially by those GSM operators who do not have UMTS spectrum.

Multi-carrier CDMA is designed to be deployed together with IS-95 in the same frequency band. The name for the multi-carrier CDMA comes from the downlink transmission direction, where instead of a single wideband carrier, multiple (up to 12) parallel narrow-band CDMA carriers are transmitted from each base station. The bandwidth is each narrow-band carrier is equal to the bandwidth of IS-95. In the uplink direction transmission is single carrier only.

Currently it seems that market is not taking off for the multi-carrier CDMA as defined as part of the IMT-2000 and developments and commercial applications are focusing on the narrowband evolution of IS-95, known as 1 × . The key operating mode of the multi-carrier CDMA was denoted as 3 × indicating 3 times larger bandwidth.

The expected frequency bands and geographical areas where these different air interfaces are likely to be applied are shown in Figure 14.1.

The WCDMA technology presented in this section is covered in much more details in [1].

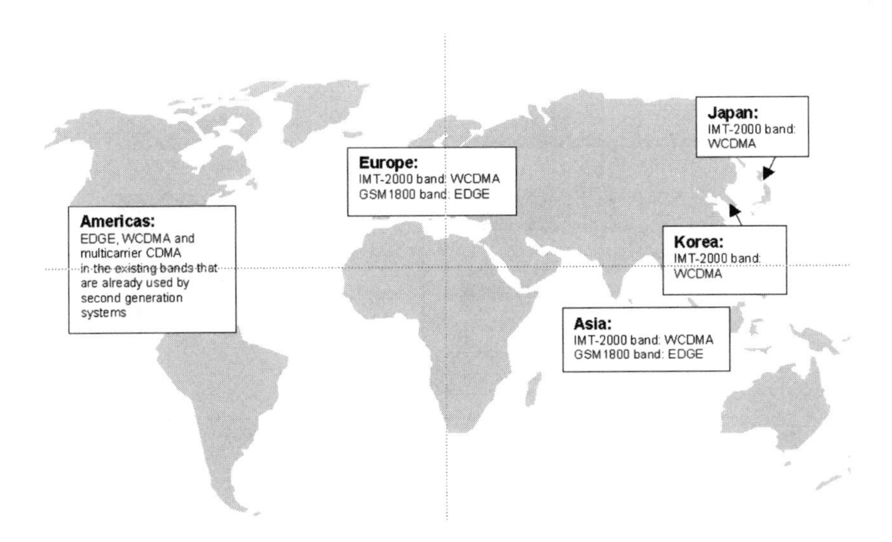

Figure 14.1 Expected air interfaces and spectrums for providing third generation services.

14.2 Spectrum for UMTS

Work to develop third generation mobile systems started when the WARC (World Administrative Radio Conference) at its 1992 meeting, identified the frequencies around 2 GHz that were available for use by future third generation mobile systems. The spectrum allocation in Europe, Japan, Korea and the US is shown in Figure 14.2. In Europe and in most of Asia the IMT-2000 bands of 2 × 60 MHz (1920–1980 plus 2110–2170 MHz) will be available for WCDMA FDD (Frequency Division Duplex). The availability of the TDD (Time Division Duplex) spectrum varies: in Europe it is expected that 25 MHz will be available for licensed TDD use in the 1900–1920 and 2020–2025 MHz bands. The rest of the unpaired spectrum is expected to be used for unlicensed TDD applications (SPA: Self Provided Applications) in the 2010–2020 MHz band. FDD systems use different frequency bands for uplink and for downlink, separated by the duplex distance, while TDD systems utilise the same frequency for both uplink and downlink.

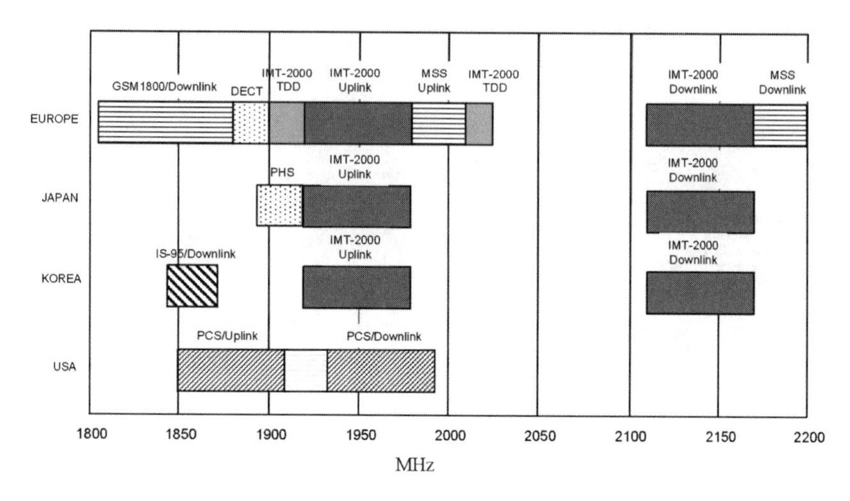

Figure 14.2 Spectrum allocation in Europe, Japan, Korea and US.

Some example UMTS license configurations are shown in Table 14.1 in Japan and in those European countries where the licenses were awarded among the first ones. The number of UMTS operators per country varies between 4 and 6.

The UMTS licenses were given either based on the operator's capabilities for providing services (beauty contest) or based on auction. From the countries listed, Finland, Japan and Spain used beauty contest while UK, Germany, Netherlands and Italy applied the auction approach. In some countries there was instead of an

Table 14.1 Example UMTS license configurations

Country	Number of operators	Number of FDD carriers (2 × 5 MHz) per operator	Number of TDD carriers (1 × 5 MHz) per operator
Finland	4	3	1
Japan	3	4	0
Spain	4	3	1
UK	5	2–3	0–1
Germany	6	2	0–1
Netherlands	5	2–3	0–1

auction, a fixed fee (not a nominal one) defined beforehand and then operators were asked to submit their applications for a license. The licensing process has been finalised in many countries but is not full complete. In some countries first phase license may have been awarded only for example 2 operators and further licenses are going to be awarded later.

More frequencies have been identified for IMT-2000. At the WARC-2000 meeting of the ITU in May 2000 the following frequency bands were identified for IMT-2000 use:

- 1710–1885 MHz
- 2500–2690 MHz
- 806–960 MHz

It is worth noting that some of the bands listed, especially below 2 GHz are partly used with systems like GSM. What shall be the exact duplexing arrangements etc. is under discussion at the moment.

14.3 WCDMA basics

The WCDMA air interface is based on CDMA technology. All the users share the same carrier with the shared resource being the power on the carrier. The characteristic feature is the wide 5 MHz carrier bandwidth over which the signal for each user is spread, as illustrated in Figure 14.3. The transmission bandwidth is the same for all data rates with the processing gain being larger for the smaller data rates

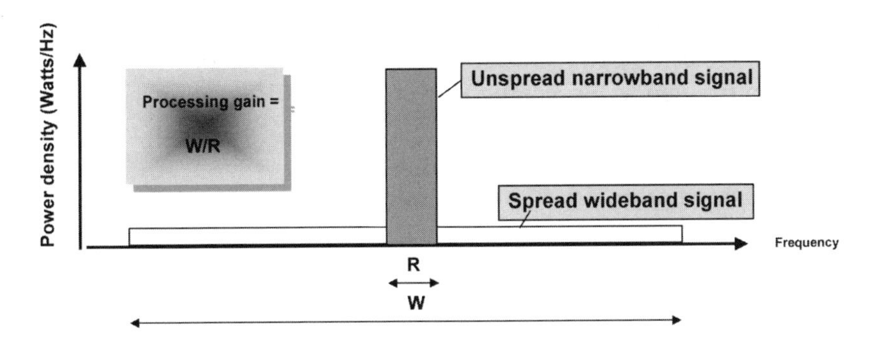

Figure 14.3 The CDMA principle used in WCDMA.

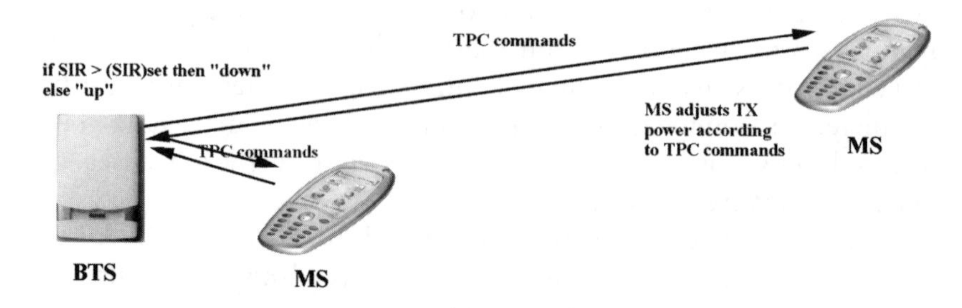

Figure 14.4 Fast power control principle.

that for the higher data rates. The achieved processing gain provides protection against the interference from the other users active on the same carrier. In the receiver despreading separates the transmitted and spread signal for data detection.

There are two key features resulting from the use of CDMA technology, namely fast power control and soft handover. They both contribute to the WCDMA system capacity but are also requirement for the proper system operation. The fast power control, especially in the uplink, is required so that users do not generate extra interference and do not block the reception of the signals from other users (Figure 14.4). Without power control a terminal transmitting near the base station would block the reception of the other users further away when exceeding the achieved processing gain. This is known also as the near-far problem in CDMA. The power control command rate in WCDMA is set to 1500 Hz with typical 1 dB step either up or down.

The soft handover is required due similar reasons. In soft handover a terminal is being connected simultaneously to two or more cells on the same frequency. Especially in the uplink this is again vital since otherwise terminal between two cells could cause problems to the cell which it is not connected. In soft handover all cells provide power control information to the terminal and the near-far problem is avoided (Figure 14.5).

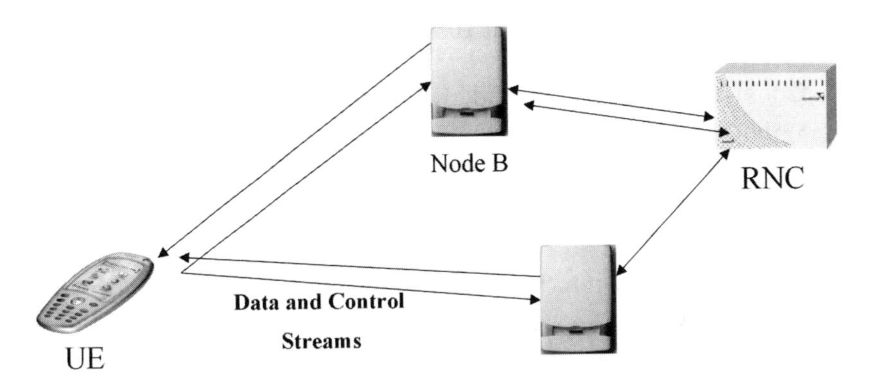

Figure 14.5 Soft handover.

14.4 *WCDMA physical layer and terminal capabilities*

In this section the main differences between the third and second generation air interfaces are described as well as the key physical layer parameters of WCDMA are explained. Also the terminal capabilities from the air interface point of view are addressed how they impact what services each type of terminal can support.

GSM and IS-95 (the standard for cdma2000 systems) are the second generation air interfaces considered here based on TDMA and CDMA respectively. Other second generation air interfaces are PDC in Japan and US-TDMA mainly in the Americas; these are based on TDMA and have more similarities with GSM than with IS-95. The second generation systems were built mainly to provide speech services in macro cells. To understand the background to the differences between second and third generations systems, we need to look at the new requirements of the third generation systems which are listed below:

- Bit rates up to 2 Mbps
- Variable bit rate to offer bandwidth on demand
- Multiplexing of services with different quality requirements on a single connection, e.g. speech, video and packet data
- Varying delay requirements

- Quality requirements from 10% frame error rate to 10–6 bit error rate
- Coexistence of second and third generation systems and inter-system handovers
- Support of asymmetric uplink and downlink traffic
- High spectrum efficiency
- Coexistence of FDD and TDD modes

The WCDMA air interface has been defined to provide in the first phase data rates up to 2 Mbps in Release – 99 and Release 4. For the Release 5 higher (peak) data rates up to 10 Mbps are expected to be part of the HSDPA (High Speed Downlink Packer Access) feature for Release 5 WCDMA standard due 03/2002.

The WCDMA way of sharing resources with the main resource being the power shared amongst users the key when dealing with variable bit rate services. When there are no fixed allocation of the physical resources that would be used by terminal even when there is not data to transmit, the capacity sharing on the carrier can be done very fast, even on 10 ms basis. The power based resource sharing is fully valid in the uplink but for the downlink more attention needs to be paid for the code resource usage when the peak and average data rates differ. For this methods have been defined for efficient use of downlink resources for example with shared channel concept for downlink packet data. The fundamental principle of WCDMA resource sharing is illustrated in Figure 14.6.

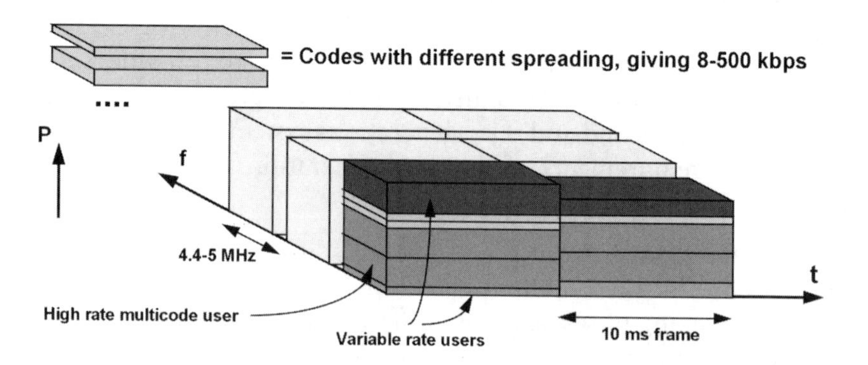

Figure 14.6 Dynamic resource sharing with WCDMA.

For other service requirements, such as different delay requirements, WCDMA provides from the physical layer point of view different Transmission Time Intervals, the value being configurable for 10, 20, 40 or 80 ms. The multi-service capability is provided already for fairly simple terminals by providing means to multiplex different services with different QoS (Quality of Service) requirements on the same connection. The simplified WCDMA service multiplexing principle is illustrated in Figure 14.7. In the actual multiplexing chain there are additional steps such as rate matching or interleaving functions that are not shown in the figure.

The QoS parameters for a particular service are mapped to the WCDMA radio interface. The QoS impacts the mode of operation

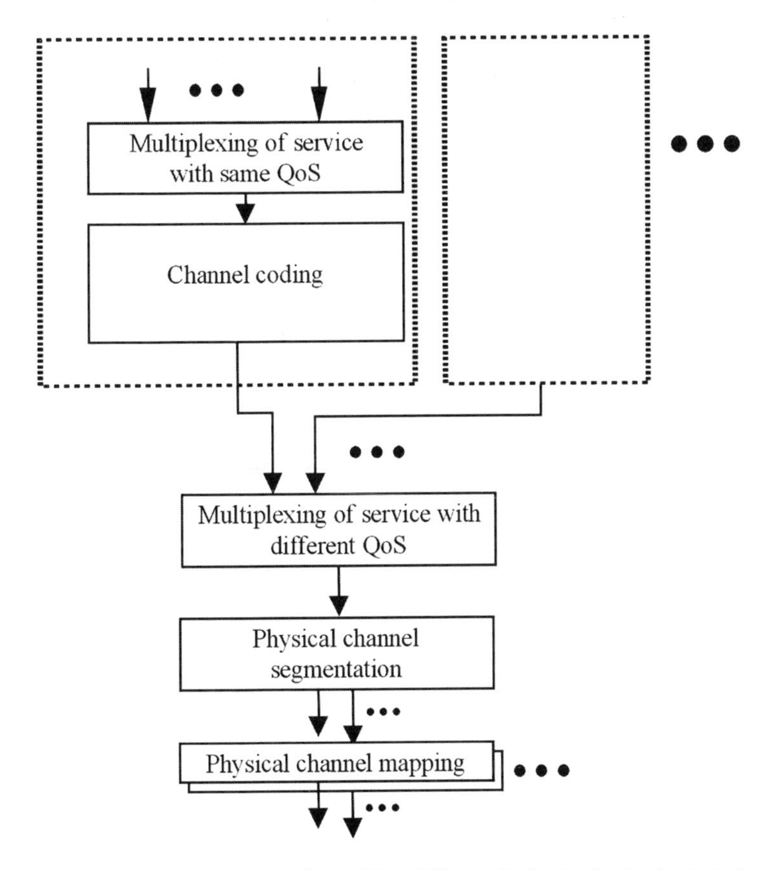

Figure 14.7 WCDMA principle of providing different QoS of a single physical channel.

in terms whether re-transmission may be used and what kind of packet error rate and probability of undetected errors are tolerated. In general the tighter the QoS parameters are, the more radio capacity is needed for the service provision. This holds with WCDMA as well as with any other radio access technology as well. What kind of channel is used to provide the service is also influenced by the terminal capability as all the features in the specification are not going to be supported by all terminals.

For the co-existence with GSM, the handover as well as necessary measurements have been specified to allow the continue the GSM service in WCDMA or vice-versa for the service set specified in GSM side.

WCDMA can provide uplink and downlink data rates independently of each other, thus facilitating uplink and downlink asymmetry per connection basis. With the variable duplex spacing it is also possible to configure less spectrum for uplink if felt necessary especially with some of the extensions bands considered to be available in the future. As the current IMT-2000 frequency allocation provides also unpaired spectrum, TDD mode of operation has been specified where users share the same carrier both uplink and downlink transmission direction. There TDMA principle has been combined with CDMA with users sharing the time slot with spreading codes as illustrated in Figure 14.8. Each time slot can be allocated either for uplink or downlink direction with some limitations. Also the wideband TDD mode of operation uses the 5 MHz bandwidth. The same solutions as with FDD mode has been applied where applicable for the TDD mode as well for maximal commonality. As part of Release 4 version of the

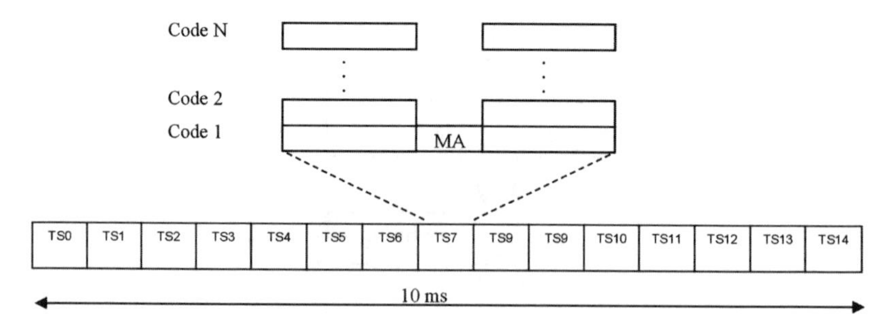

Figure 14.8 WCDMA TDD mode of operation with TDMA/CDMA principle applied.

specifications, also a narrowband TDD mode has been introduced with a chip rate 1.28 Mcps compared to 3.84 Mcps for FDD and TDD in Release '99. Currently it seems that market is now focusing on FDD only and TDD deployment is to be considered at a later stage.

To achieve high spectrum efficiency besides diversity methods due to the bandwidth, WCDMA also incorporates methods not used in 2G standards such as transmit diversity in the downlink direction. Also the advanced channel coding methods provide coding gain to achieve low SIR (Signal to Interference Ratio) requirements for the desired quality level.

In WCDMA the same principle as with GSM with Terminal class mark is not used. WCDMA terminals shall tell the network upon connection set up large set of parameters which indicate the capabilities of the particular terminal. It is worth noting that terminal capabilities can be given independently for the uplink and downlink direction. To provide guidance which capabilities should be applied together, example capability classes have been specified in WCDMA standardisation in 3GPP (3rd Generation Partnership Project). The following capability classes have been defined for WCDMA:

- 32 kbps class: This is intended to provide basic speech service, including AMR speech as well as some data limited data rate capabilities but not together with speech service.
- 64 kbps class: This is intended to provide speech and data service, with also simultaneous data and speech capability e.g. AMR speech and 32 kbps data simultaneously.
- 128 kbps class: This class has the air interface capability to provide for example video telephony or then various other data services.
- 384 kbps class is being further enhanced from 144 kbps and has for example multicode capability which points toward support of advanced packet data methods provided in WCDMA. This class can provide e.g. AMR speech simultaneously with packet data up to 384 kbps. From Release 4 onwards terminals with 384 kpbs or higher data rate capability will support advanced downlink packet data operation features.
- 768 kbps class has been defined as an intermediate step between 384 kbps and 2 Mbps class.

- 2 Mbps class: This is the state of the art class and the 2 Mbps capability has been defined for downlink direction only. (Note that 2 Mbps can still be provided in the uplink as well).

The intention is also that each example class can provide the services that are provided by the classes with lower data rates capabilities. A terminal can provide the capability with FDD or then with both FDD and TDD. It is from the capability point of view also possible to make TDD only terminal, whether such devices will be seen in the market place is another question.

14.5 WCDMA air interface performance

From the end user perspective the quality of cellular service may be limited by the weak signal, i.e. by the coverage, or by too many users trying the access the service at the same time, i.e. by the capacity of the system. The coverage and capacity of WCDMA is covered in this section.

Coverage is important when the network is not limited by capacity, such as at the time of initial network deployment, and typically in rural areas. Macro cell coverage is determined by the uplink range from the mobile station to the base station, because the transmission power of the mobile is much lower than that of the macro base station. The output power of the mobile is typically 21 dBm (125 mW) and that of the macro cell base station 40–46 dBm (10–40 W) per sector.

The maximum cell size depends on the required bit rate: the higher the bit rate, the smaller is the cell area. Typically, about 40% more sites are needed if the available bit rate at the cell edge need to be doubled, e.g., from 32 to 64 kbps.

To understand the typical cell areas of WCDMA cells, we can compare the WCDMA cell area to the existing GSM cell areas. That comparison is also important since GSM operators will utilise their existing cell sites also for their UMTS networks. With existing GSM1800 sites WCDMA offers bit rates of 64–144 kbps with the same coverage probability as for GSM1800 speech. With GSM900 sites WCDMA offers bit rates of about 32 kbps with the same coverage probability as for GSM900 speech assuming that simple smart

20 Vignettes from a 3G Future

Click for more, click to talk, click to buy,

I really have to admire the way the UMTS services think of me by giving me direct links to relevant additional information and services. I still remember how cool it was when on the Internet you found a "link" and you could click on it to move to another location or another page with further information. On the UMTS phone now, that idea has moved much further. There is click for almost anything. Of course there is click to get more information, but there are special click to talk buttons, so if I want to talk to a real person about some information, often there is a click to talk button and I am instantly connected. The other is click to buy, which I really appreciate.

UMTS operators can build more revenues out of click-to-talk where a voice connection is added to a live data connection. The voice call mostly would be at a higher cost than the data browsing going on, bringing more revenues. Click to buy allows for instant approval to act on a purchasing impulse, helping get mobile commerce into the mass market.

antenna solutions are used in WCDMA. The cell size in GSM900 can be larger than in GSM1800 because the signal attenuation is lower in lower frequency. Therefore, the coverage of WCDMA in 2 GHz band will not be as good as for GSM900 speech in 900 MHz band without smart antenna solutions. On the other hand, the GSM site density is in most urban areas denser to provide more capacity, and therefore, reusing existing GSM sites will provide a high quality WCDMA network.

The typical site areas for GSM1800 and for WCDMA will be 2 km^2 in urban area, 5–10 and 20 km^2 in rural area. A 3-sector site is assumed in here. The exact maximum site area depends on the environment and on the required coverage probability.

When the traffic increases in the network, a higher base station density may be needed. The typical capacity of WCDMA air interface is 1 Mbps per carrier per sector per 5 MHz carrier in 3-sector macro cells and 1.5 Mbps micro cell per 5 MHz carrier. This value represents a maximum total throughput per sector which can be shared between the users who request service at the same time. It is the radio resource management algorithms in the radio network controller that allocate the capacity between the users. The capacity allocation can be based on the requested services and their priorities.

Providing high capacity will be challenging in urban area where the offered amount of traffic can be very high. If we assume first that third generation services are provided using 1-carrier macro cells with 2 km^2 site area, the available capacity will be 1.5 Mbps/km^2/carrier. If the operator has three carriers, the capacity will be 4.5 Mbps/km^2. When more capacity is needed, an operator can add more macro sites, or to deploy micro cells. Let's assume that the maximum site density is 30 micro sites per km^2. Using three carriers the micro cell layer would offer $30 * 1.5 * 3 = 135$ Mbps/operator/km^2. If we look at the total capacity of the UMTS band of 12 carriers, it could be up to 500 Mbps/km^2 if all operators deploy micro cells. Using indoor pico cells with FDD or with TDD can further enhance the capacity. TDD is most suited for small cell deployment due smaller achievable cell radius. The coverage areas mentioned earlier are valid for FDD only. The figures presented do not include the potential of the WCDMA enhancements currently being finalised in 3GPP, such as HSDPA. HSDPA results from the 3GPP feasibility study have indicated possibility to double the capacity with downlink packet data operation. Also for any comparison with other radio access technologies the support of e.g. transmit diversity in all WCDMA terminals gives WCDMA competitive advantage in the market place (Table 14.2).

Table 14.2 Typical capacities of WCDMA air interface

	Macro cell layer	Micro cell layer
Capacity per site per carrier	3 Mbps with 3 sectors	1.5 Mbps
Capacity per site per operator with 3 UMTS FDD carriers	9 Mbps	4.5 Mbps
Maximum site density in urban area	5 sites/km^2	30 sites/km^2
Maximum capacity per operator	45 Mbps/km^2	135 Mbps/km^2

HSDPA improves WCDMA air interface performance over Rel'99/R4. The main gains for the system performance are summarised as follows

1 Improved spectral efficiency in all environments, typically 50–100% higher average cell throughput than with Rel'99/R4. This is achieved with hybrid ARQ and adaptive modulation and coding.
2 Better support for low delay services with faster retransmission: streaming QoS can be supported on HSDPA and TCP/IP (Transmission Control Protocol Internet Protocol) works more fluently with faster retransmissions.
3 Higher peak bit rates exceeding 10 Mbps are supported from the physical layer point of view with adaptive modulation and coding.
4 HSDPA co-exists on the same carrier as Rel'99/R4 WCDMA and Rel'99/R4 UEs. HSDPA can also shared physical connection with conversational services like voice and video.
5 Future proof concept. HSDPA can take advantage of high SIR values. SIR can be improved with future receiver/transmitter concept, like improved baseband processing (Table 14.3, Figure 14.9).

Table 14.3 Maximum HSDPA bit rates ([1]64QAM is not foreseen to be part of Release 5 specifications.)

Modulation	10 codes with SF = 16	15 codes with SF = 6
QPSK	3.6 Mbps	5.4 Mbps
16QAM	7.2 Mbps	10.8 Mbps
64QAM[a]	10.8 Mbps	16.2 Mbps

[1] 64QAM is not foreseen to be part of Release 5 specifications.

WCDMA is designed to be deployed together with GSM network, and handovers are supported from WCDMA to GSM and from GSM to WCDMA. In the initial deployment phase the WCDMA network coverage will not be as large as GSM network coverage and handovers from WCDMA to GSM can be used to provide continuous service even if the user leaves WCDMA coverage area in connected mode. Also idle mode parameters are supported for inter-system cell reselection.

The inter-system handovers are triggered by the source radio access network, and the handover triggers are vendor specific. Inter-system handover can be triggered also due to high load to balance loading between WCDMA and GSM to fully utilise GSM and WCDMA networks together.

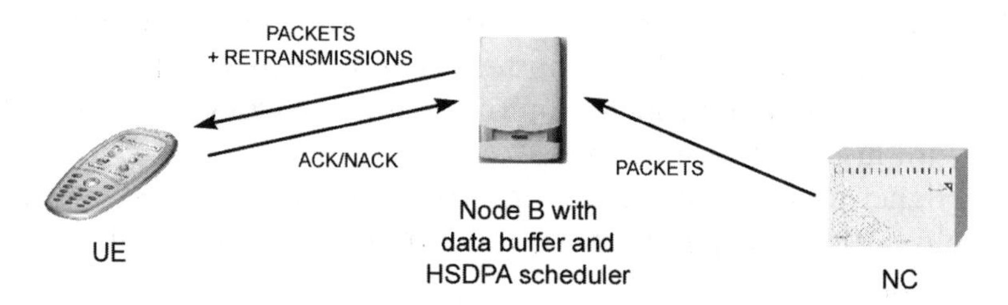

Figure 14.9 HSDPA implications to Node B functionality.

14.6 *In summary*

This chapter covered briefly the air interfaces, the spectrum, WCDMA basics physical layers and terminal capabilities, as well as the WCDMA performance. WCDMA technology offers wide range of possibilities for service introduction, including means for providing different quality of service as well as means for providing multiple services simultaneously. For those who want to read more about the WCDMA technology of UMTS please see the bibliography section of this book. And for those who ponder if there will be too much capacity, we are reminded of Governor Jerry Brown's famous quotation: "Too often I find that the volume of paper expands to fill the available briefcases."

References

[1] H. Holma & A. Toskala, 'WCDMA for UMTS', Revised Edition, John Wiley & Sons, 2001.

15

'Yesterday I was a dog. Today I am a dog. Tomorrow I will still be a dog. Sigh. There is so little room for advancement.'

Snoopy

Postscript:

Final Thoughts

Tomi T Ahonen and *Joe Barrett*

The mobile phone has become part of everybody's personal outfit, taken everywhere and kept close at all times. This connectedness has brought about a new immediacy and the ability to 'always communicate'. The services delivered by UMTS (Universal Mobile Telecommunications System) will greatly enhance this customer experience creating a richer communication environment.

In this world where multiple connectedness has become the norm it is even more important to be able 'to network' or to build personal networks and manage those relationships in new and efficient ways. When the Industrial Age was turning into the Information Age, information and control of it became increasingly valuable. When information was collected and processed, but not delivered for further handling, *controlling* information was power. In the late middle ages the cardinals and other religious leaders were the epitome of this type of structure as they were the only ones able to read Latin and thus controlled information flow. The last vestiges of those who still believe in hoarding and controlling information are now being totally destroyed.

The Information Age is now changing into the Connected Age; the age of the converged UMTS Mobile Internet. In the Connected Age *sharing* information is power. It is the latest evolution of the Information Age, the Computer Age, the Networked Computer Age, the Mobile Age, the Converged Age. Where we stand now there is the traditional fixed Internet and the new UMTS Mobile Internet. These two overlap, but eventually there will be more connected devices in the UMTS mobile world than on the fixed Internet.

When the majority of web surfers access content from their mobile terminals, that content will migrate to serve those mobile surfers more efficiently. That transition is inevitable. It is likely that the content providers will recognise new opportunities to make *more money* in the UMTS Mobile Internet than on the fixed Internet. These content providers may prefer this new mobile domain and migrate their services to UMTS much faster than we now expect. The question is only 'When?' We believe it will definitely happen sooner than many currently industry observers think.

As a content provider might view it, if one media channel has more users and generates money more easily than its competitor media channel, it is clear where the focus on content improvement will lie. The best content will soon be found on the Mobile Internet. In this book we have shown a way to take any existing digital content, and with the 5 M's (Movement Moment Me Money Machines), make the content more relevant, more useful, and more profitable in its Mobile Internet form.

We believe it is inevitable that the majority of users, the best content, and most of the money will migrate to the Mobile Internet. When that has happened we will recognise that the UMTS Mobile Internet will have triumphed over the fixed Internet. Perhaps then we will have a more natural sounding term for the UMTS Mobile Internet, something like 'mobilenet' or the 'wireless web' which could be called 'WW' or 'W2' to distinguish the discussion from the (old fashioned and fixed) Internet and the WWW (World-Wide Web).

At the centre of the UMTS mobile domain will be the mobile network operators who control access and information flow to and from the new data enabled mobile terminals. The real positions of strength in the new UMTS world will come out of the control of the access interface. That position is at least initially with the UMTS network operators. But they cannot create *all* the content and utility for their networks. Even the biggest global UMTS network operators have to network with their partners and at times cooperate with their competitors, thus sharing and creating the key to success.

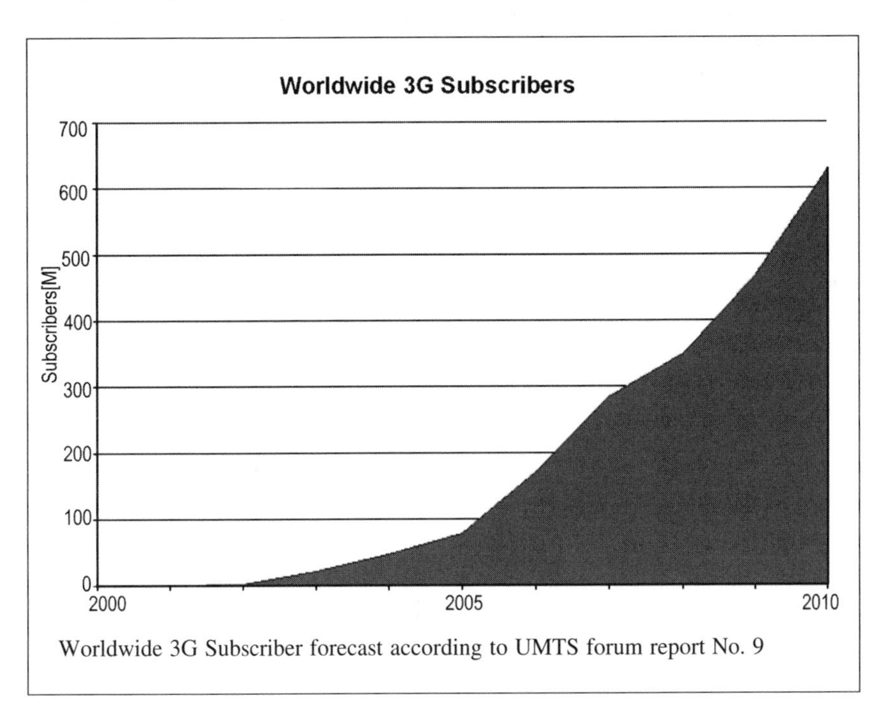

Worldwide 3G Subscriber forecast according to UMTS forum report No. 9

Operators will need to partner and those that resist or look to overtly control all their relationships will become less successful. Hopefully some of the services outlined in this book have helped identify what kind of services could be created, how they can be marketed, and to give some insight into who are natural partners in those situations. We also hope that this book stimulates thinking about the revenue sharing and profit aspects of creating new Mobile Internet services.

This book is an early attempt to look at some of the services we believe will emerge in the UMTS environment. We took a look at different areas of the services, listed over 250 services and discussed in some detail over 200 services. We also analysed parts of the business case value chains involved. This book has discussed the services for

UMTS which is where 'serious' money will be made over the first decade of the new millennium. A trillion dollar's worth of it per year before the first decade is done. Why write a book about new mobile services? We are reminded of the famous bank robber Willie Sutton's reply to why he robbed banks: "Because that's where the money was."

We hope that this book has inspired you.

Abbreviations

0–1–2–3	0 manuals, 1 button internet, 2 seconds maximum delay, 3 keystroke maximum
1xRTT	Single carrier Radio Transmission Technology (IS–95 evolution)
2G	2nd Generation mobile networks
3D	Three Dimensional
3G	Third Generation mobile systems
3GPP	3rd Generation Partnership Project
4 P's	(Service Creation Aid): Portable Personal Processor Proactive
5 M's	Movement Moment Me Money Machines
64QAM	64–Quadrature Amplitude Modulation
ADSL	Asynchronous Digital Subscriber Line
AIDA	Awareness, Interest, Desire, Action
AMR	Adaptive MultiRate
AOL	America On–Line
API	Application Programmable Interface
APPSE	APPlication SErver
ARPHU	Average Revenue Per Human User
ARPMU	Average Revenue Per Machine User
ARPU	Average Revenue Per User
ARQ	Automatic Repeat Request (retransmission)
ASI	Access and Share Information
ASP	Application Service Provider
ATM	Automatic Teller Machine

B2B	Business to Business
B2C	Business to Consumer
B2E	Business to Employee
BTS	Base Trasceiver Station
CAPEX	CAPital EXpenses
CCB	Customer Care & Billing
CD	Compact Disk
CDMA	Code Division Multiple Access
CDR	Call Detail Record
CLI	Calling Line Identification
CLIP	Calling Line Identification Presentation
CLP	Calling Line Picture
CPS	Call Processing Server
CRM	Customer Relationship Management
CSCF	Call State Control Function
DBm	dB compared to milli, i.e. 1000 times higher value
DECT	Digital Enhanced Cordless Telephone
DoF	Department of the Future
DVD	Digital Video Disc
EBITDA	Earnings Before Interest, Taxes, Depreciation and Amortization
ECG	ElectroCardioGram
EDGE	Enhanced Data rates for GSM Evolution
E–OTD	Enhanced Observed Time Difference
ERP	Enterprise Resource Planning
ETSI	European Telecommunications Standards Institute
FDD	Frequency Division Duplex
FOMA	Freedom Of Multimedia Access
FTSE 100	Financial Time Stock Exchange 100
GDP	Gross Domestic Product
GPRS	General Packet Radio System
GSM	Global System for Mobile communications
HSCSD	High Speed Circuit Switched Data
HSDPA	High Speed Downlink Packer Access
HTML	HyperText Markup Language
Hz	Herz
ID	Identity
IETF	Internet Engineering Task Force

IM	Information Management
IMPEX	IMPlementation EXpense
IMT–2000	International Mobile Telephony (2000)
IN	Intelligent Network
IP	Internet Protocol
IPv4	Internet Protocol version 4
IPv6	Internet Protocol version 6
IS–95	CDMA One
ISDN	Integrated Services Digital Network
ISP	Internet Service Provider
IT	Information Technology
ITU	International Telecommunications Union
kbps	kilobits per second
KISS	Keep It Simple, Stupid
km	kilometre
MA	MidAmble
mAd	Mobile Advertising
MAGIC	Mobile Anytime Globally Integrated Customized
Mbit/s	Megabits per second
Mbps	Megabits per second
MDF	Master Design Framework
MHz	MegaHerz
MIME	Multipurpose Internet Mail Extensions
MMS	Multimedia MeSsaging
MNO	Mobile Network Operator
MP3	MPEG audio Layer 3
MP4	MPEG structured audio 4
MPEG	Motion Picture Experts Group
MS	Mobile Station
MTV	Music TV
MVNO	Mobile Virtual Network Operator
NTT	Nippon Telephone and Telegraph
mW	milliWatts
OPEX	OPerating EXpenses
OSA	Open Service Architecture
OSS	Operational Support System
PAIR	Personal Available Immediate Real time
PBX	Private Branch eXchange

PC	Personal Computer
PCS	Personal Communication Systems
PDA	Personal Digital Assistant
PDC	Personal Digital Cellular
PHS	Personal Handy phone System
PIM	Personal Information Manager
PIN	Personal Identity Number
PKI	Public Key Infrastructure
POTS	Plain Old Telephone System
POV	Plain Old Voice
PSTN	Public Switched Telephone Network
PTT	Post, Telephone and Telegraph
QoS	Quality of Service
R&D	Research and Development
RAN	Radio Access Network
Rel	Release
Rel99	3GPP standard released 03/00
RelR4	3GPP standard released 03/01
SCM	Supply Chain Management
SDP	Session Description Protocol
SI	System Integration
SIM	Subscriber Identity Module
SIP	Session Initiation Protocol
SIR	Signal to Interference Ratio
SKU	Stock Keeping Unit
SME	Small and Medium Enterprise
SMS	Short Message Service
SOHO	Small Office Home Office
SPA	Self Provided Applications
SWIM	See What I Mean
SyncML	Synchronising Markup Language
TCP/IP	Transmission Control Protocol Internet Protocol
TDD	Time Division Duplex
TDMA	Time Division Multiple Access
TIM	Telecom Italia Mobile
TPC	Transmission Power Control
UE	User equipment
UI	User Interface

UMTS	Universal Mobile Telecommunications System (also Universal Mobile Telecommunication Services)
URL	Universal Resource Locator
US–TDMA	United States Time Divisional Multiple Access
VAS	Value Added Services
VCR	Video Cassette Recorder
VHE	Virtual Home Environment
VNO	Virtual Network Operator
VoIP	Voice over IP (Internet Protocol)
VPN	Virtual Private Network
W	Watt
WAA	Wireless Advertising Association
WAP	Wireless Application Protocol
WARC	World Administrative Radio Conference
WCDMA	Wideband Code Division Multiple Access
W–LAN	Wireless Local Area Network
WTA	Wireless Telephony Application
WWW	World–Wide Web

Bibliography

Black, U. ATM, volume II: Signaling in broadband networks. Englewood Cliffs, NJ: Prentice Hall, 1998, 224 pp.

Black, U. ATM, volume I: Foundation for broadband networks. Englewood Cliffs, NJ: Prentice Hall, 1999, 450 pp.

Comer DE. Internetworking with TCP/IP, volume I: Principles, protocols and architecture. 4th edition. Englewood Cliffs, NJ: Prentice Hall, 2000, 755 pp.

Doraswamy N, Harkins 0. IPSec: the new security standard for the internet, intranets and virtual private networks. Englewood Cliffs, NJ: Prentice Hall, 1999, 216 pp.

Ericsson Telecom AB. Understanding telecommunications 1–2. Studentlitteratur AB, 1997, 493 + 677 pp

Glisic 5, Vicetic B. Spread spectrum CDMA systems for wireless communications. Boston, MA: A–H Publishers, 1997, 383 pp

Händel R, Huber MN, Schröder S. ATM networks, concepts, protocols, applications, second edition. Reading, MA: Addison–Wesley, 1994, 287 pp

Hannula I, Linturi R. 100 phenomena, virtual Helsinki and the cybermole (translation from Finnish by William More) e–book at http://www.linturi.fi/100_phenomenal Helsinki: Yritysmikrot 1998 212 pp

Heine G. GSM Networks: protocols, terminology and implementation. Boston, MA: Artech House, 1998. 416 pp

Holma H, Toskala A. WCDMA for UMTS, revised edition. Chichester: Wiley, 2001, 313 pp

IDATE. Web Music: Issues at Stake and Forecasts. Montpellier 2001

Kaaranen H, Ahtiainen A, Laitinen L, Nahgian S, Niemi V. UMTS networks. Chichester: Wiley, 2001, 302 pp

Kopomaa, T. City in your pocket, the birth of the information society. Helsinki: Gaudeamus, 2000, 143 pp

Laiho J, Wacker A, NovQsad T. Radio network planning and optimisation for UMTS. Chichester: Wiley, 2001, 512 pp

Lee, WCY. Mobile cellular telecommunications; analog and digital systems, second edition, New York: McGraw–Hill, 1995, 664 pp

Materfield R. Telecommunications signalling. UK: The Institution of Electrical Engineers, 1999. 435 pp.

Menezes AJ, van Oorschot PC, Vanstone S. Handbook of applied cryptography. Boca Raton, FL: CRC Press, 1996, 760 pp

Mouly M, Pautet M–B. The GSM system for mobile communications. France: Mouly M, Pautet, MB, 1992, 701 pp

Ojanperä T, Prasad R. Wideband CDMA for third generation mobile communications. Boston, MA: Artech House, 1998, 439 pp

Rantalainen T., Spirito MA, Ruutu V. Evolution of location services in GSM and UMTS networks. Proceedings of the third international symposium on wireless personal multimedia communications (WPMC 2000), November 2000, Bangkok, pp 1027 – 1032

Redl SH, Weber MK, Malcolm WH. An introduction to GSM. Boston MA: Artech House, 1995, 379 pp

Spirito MA. Mobile stations location estimation in current and future TDMA mobile communication systems. Ph.D. Thesis Politecnico di Torino, Facolta di Ingegneria, 2000

Stallings W. Data & computer communications. Englewood Cliffs, NJ: Prentice Hall, 2000, 810 pp

Steffens J. Newgames strategic competition in PC revolution. NY: Pergamon 1993, 220 pp

Tennant J, Friend G, Economist guide to business modelling, London: Economist Books 2001, 272 pp

UMTS Forum, Report 16: 3G portal study, available at the UMTS Forum website http://www.umts–forum.org/reports_r.html November 2001

UMTS Forum, Report 13: Structuring the service opportunity, available at the UMTS Forum website *http:/Iwww.* umts–forum.org/reports_r.html April 2001

UMTS Forum, Report 12: Naming, addressing and identification issues for UMTS, available at the UMTS Forum website http://www. umts–forum.org/reports_r.html February 2001

UMTS Forum, Report 11: Enabling UMTS third generation services and applications, available at the UMTS Forum website http://www.umts–forum.org/reports_r.html October 2000

UMTS Forum, Report 10: Shaping the mobile multimedia future, available at the UMTS Forum website http://www.umts–forum.org/reports_r.html October 2000

UMTS Forum, Report 9: UMTS third generation market structuring the service revenue opportunities, available at the UMTS Forum website http://www.umts–forum.org/reports_r.html October 2000

Witerbi AJ. CDMA principles of spread spectrum communications. Reading, MA: Addison–Wesley, 1995, 254 pp

Useful Websites

160 characters (SMS and Mobile Messaging Association)
http://www.160characters.com/

3GPP (Third Generation Partnership Project)
http://www.3gpp.org/

ALACEL (Latin American Wireless Industry Association)
http://www.alacel.com/home.cfm?Iang=en

ARIB (Association of Radio Industries and Business) (Japan)
http://www.arib.or.jp/index_English.html

Baskerville Publications
http://www.baskerville.telecoms.com

CWTS (China Wireless Telecommunication Standards Group)
http://www.cwts.org/cwts/index_eng. html

CTIA (Cellular Telecommunications & Internet Association)
http://www.wow-com.com/

GSA (Global mobile Suppliers Association)
http://www.gsacom.com/

GSM Association
http://www.gsmworld.com

ETSI (European Telecommunications Standards Institute)
http://www.etsi.org/

FCC (Federal Communications Commission)
http://www.fcc.gov/

IDATE
http://www.idate.fr

IEEE (Institute of Electrical and Electronics Engineers)
http://www.ieee.org/

IETF (Internet Engineering Task Force)
http://www.ietf.org

ITU (International Telecommunications Union)
http://www.itu.ch/

MDA (Mobile Data Association)
http://www.mda-mobiledata.org/

MEF (Mobile Entertainment Forum)
http://www.mobileentertainmentforum.org/

MGIF (Mobile Gaming Interoperability Forum)
http://www.mgif.org!

OFTEL (Office of Telecommunications) (UK)
http://www.oftel.gov.uk/

PCIA (Personal Communications Industry Association)
http://www.pcia.com/

SyncML
http://www.syncml.org/

TI (Committee TI)
http://www.t1.org/

TTA(Texas Telephone Association)
http://www.tta.org/

TTC (Telecommunication Technology Committee) (Japan)
http://www.ttc.or.jpte/

UMTS Forum
http://www.umts-forum.org/

UWCC (Universal Wireless Communications Consortium)
http://www.uwcc.org/

WAA (Wireless Advertising Association)
http://www.waaglobal.org/

WAP Forum
http://www.wapforum.org

WCA (Wireless Communications Association International)
http://www.wcai.com/

WISPA (Wireless Internet Service Providers Association)
http://www.wispa.org/

WLANA (Wireless LAN Association)
http://www.wliaonhne.com/

WLIA (Wireless Location Industry Association)
http://www.wliaonline.com/

WMA (Wireless Marketing Association)
http://www.wirelessmarketing.org.uk/

Index to Services

Index of Key Words and Phrases

(Services In Italics)